Mathematics and
the Development of the
Physical Sciences

Lynchburg College Symposium Readings

Third Edition

2007

Volume VI
Mathematics and the
Development of the
Physical Sciences

Edited by

Kevin Peterson, PhD

To order additional copies of this book, contact:
Xlibris Corporation
1-888-795-4274
www.Xlibris.com
Orders@Xlibris.com
23031

CONTENTS

III. Science and Mathematics of the Twentieth Century

ACKNOWLEDGEMENTS

Abbott, E. A. *Flatland*. Permission granted by Dover Publishing

Copernicus, Nicolaus. 1959. *Commentariolus* from *Three Copernican Treatises*. Dover Publications. Reprinted by permission of Dover Publishing.

Cedering, Siv. 1984. *Letters From a Floating Works: New and Selected Poems*. Pittsburg Press. Reprinted by permission of Siv Cedering.

Feynman, Richard P.; QED © 1985 Princeton University Press. Reprinted by permission of Princeton University Press.

From DISCOVERIES AND OPINIONS OF GALILEO by Galileo Galilei, translated by Stillman Drake, copyright © 1957 by Stillman Drake. Used by permission of Doubleday, a division of Random House, Inc.

GODEL ESCHER BACH by DOUGLAS HOFSTADTER.ISBN 0465026567. Copyright © 1979 by Basic Books, a member of Perseus Books, L.L.C.

Hardy, G. H. 1967. *A Mathematician's Apology*. Reprinted with the permission of Cambridge University Press

Maxwell, J.C. *Matter and Motion*. Dover Publications. Reprinted by permission of Dover Publications.

Newton, Isaac. Mathematical Principles of Natural Philosophy and His System of the World (Principia). Reprinted by permission of The University of California Press.

Pauling, Linus. *No More War!* Reprinted by permission of Dr. Linus Pauling, Jr.

Pauling, Linus. 1947. *General Chemistry An Introduction to Descriptive Chemistry and Modern Chemical Theory.* W. H. Freeman, Publishers. Reprinted by permission of Dr. Linus Pauling, Jr.

POLYA, G.; HOW TO SOLVE IT. © 1945. Princeton University Press, 1973 renewed PUP. Reprinted by permission of Princeton University Press.

Cartwright, M. L. 2004. "Mathematics and Thinking Mathematically." *Musings of the Masters,* R.G. Ayoub, Ed. Reprinted with permission by Mathematics Association of America.

Lynchburg College Symposium Readings Senior Symposium and the LCSR Program

The ten-volume series, Lynchburg College Symposium Readings, has been developed by Lynchburg College faculty for use in the Senior Symposium and the Lynchburg College Symposium Readings Program (SS/LCSR). Each volume presents primary source material organized around interdisciplinary, liberal arts themes.

In 1976, the College developed the Senior Symposium as a two-hour, interdisciplinary course, required of all graduating seniors. On Mondays, students in all sections of the course come together for public lectures, given by invited guest speakers. On Wednesdays, Symposium students meet in their sections for student-led discussions and presentations on associated readings from the LCSR series. The course requires students, who have spent their later college years in narrowing the scope of their studies, to expand their fields of vision within a discussion group composed of and led by their peers. Students can apply analytical and problem-solving capabilities learned in their major fields of study to issues raised by guest speakers and classical readings.

This approach works against convention in higher education, which typically emphasizes the gradual exclusion of subject areas outside of a student's major field. But Senior Symposium leads students, poised for graduation, into their post-college intellectual responsibilities. They gain experience in taking their liberal education into real world problems, using it to address contemporary problems thoughtfully and critically. In order to do this successfully, students must abandon their habitual posture as docile receptors of authoritative information for a much more skeptical attitude toward opinion, proof, reasoning, and authoritative experience. The effort to think constructively through a variety of conflicting opinions—on a weekly basis—prepares them well for the mature, independent, well-reasoned points of view expected of educated adults.

The LCSR Program's primary goals are to foster an appreciation of the connection between basic skills and interdisciplinary knowledge, and to promote greater cross-disciplinary communication among faculty and students. General education core courses or courses that serve other program requirements may be classified as "LCSR," as long as they fulfill certain speaking and writing activities connected to LCSR readings. The effect of the program has been to help create the atmosphere of a residential academic community; shared learning creates a climate in which teaching and learning take root more forcefully.

Since its inception, the SS/LCSR Program has helped create opportunities for faculty interaction across the disciplines in "pre-service" and "in-service" workshops. Each May, the LCSR Program sponsors a four-day pre-service workshop to which all new full-time and part-time faculty members are invited. Participants receive individual sets of the Lynchburg College Symposium Readings, which they make their own by using them during the workshop in exercises designed to promote familiarity with a wide variety of the readings. The goals of the workshop are several: for those unfamiliar with the program, to begin planning an LCSR course; for new faculty, to become acquainted with those they have not met and get to know their acquaintances better; for other faculty of various experiential levels, to share their pedagogical successes; for new teachers, to ask questions about teaching, about the College, and about the students in an informal setting; for experienced teachers to re-visit some of their assumptions, pedagogies, and strategies; to inspire strong scholarship, creative teaching, risk-taking, and confidence among all participants.

Another opportunity comes with the "in-service" workshops, which occur each month during the school year. The LCSR Program sponsors luncheons and dinners at which faculty teaching in the program give informal presentations on their use of specific teaching strategies or reading selections. Attendance is voluntary, but many try to be present for every session. For those involved in the LCSR program, teaching has become what Lee Schulman, President of the Carnegie Foundation, calls "community property."

On the Lynchburg College campus, there is evidence of a systematic change in teaching effectiveness as well as sustained faculty commitment to the program. By the 2002-2003 academic year, nearly two-thirds of all full-time faculty members and more than half of all part-time faculty members had completed the workshop at some time during their time at Lynchburg College. In any given semester, roughly ten to fifteen percent of the total class enrollments are in LCSR courses, not counting the required Senior Symposium. An important feature of this program is that participation is voluntary on the part of the faculty and, except for the required Senior Symposium, on the part of students. The program's influence quietly pervades the campus community, improving teaching and scholarship. Many see the LCSR Program as the College's premier academic program.

The Senior Symposium/LCSR program publishes the *Agora*, an on-line publication of Lynchburg College specializing in responses to the great books of the world.

An official publication of the Association of Core Texts and Courses (ACTC), students from colleges and universities who are institutional members may submit their work for consideration. Aiming to integrate classical ideas and issues with contemporary ones, the *Agora* takes its title from the marketplace at the heart of classical Athens, where much of Athenian public life was carried on: mercantile exchange, performance, political debate, athletic contests, and the public worship of deities, all took place within the hustle and bustle of the Athenian agora. Similarly, the journal seeks to be a marketplace for important ideas and issues.

Since 1976, the Senior Symposium and the LCSR Program have affected the academic community both within and beyond the Lynchburg College campus. In 1991, Professor Richard Marius from Harvard University's writing center in reviewing the program favorably said, "I have seldom in my life been so impressed by an innovation in college education. I suppose the highest compliment that I could pay to the program was that I wished I could teach in it." Also in 1991, Professor Donald Boileau of George Mason University's Communication Studies department wrote, "what I discovered was a sound program that not only enriches the education of students at Lynchburg College, but what I hope can be a model for many other colleges and universities throughout our country." In spring 2003, in an article titled, "Whither the Great Books," Dr. William Casement described the LCSR program as the "most fully organized version" of the recent growth of great-books programs that employ an "across-the-disciplines structure." According to him, this approach perhaps encourages the use of the great books across the curriculum, which can be less isolating than even interdisciplinary programs. The Senior Symposium and LCSR Programs have received national acclaim in such publications as Loren Pope's *Colleges That Change Lives* and Charles Sykes and Brad Miner's *The National Review of College Guide, America's 50 Top Liberal Arts Schools.*

Mathematics and the Development of the Physical Sciences is the sixth volume of the third edition of the Lynchburg College Symposium Readings series. We gratefully acknowledge the work of Dr. Julius Sigler, Professor of Physics and Vice President for Academic Affairs and Dean of the College who edited the first and second editions of this volume. The Senior Symposium was the creation of Dr. James Huston, Dean of the College, *Emeritus* (1972-1984). With Dr. Michael Santos, Professor of History, he co-founded the LCSR program. Dean Huston served as the first series editor, and with Dr. Sigler, co-edited the second series. All three remain committed to the program today, and for this we are grateful.

Peggy Pittas, PhD
Series Managing Editor
Katherine Gray, PhD
Copy Editor

INTRODUCTION

Nothing in life is to be feared, it is only to be understood.
Now is the time to understand more, so that we may fear less.
Marie Curie

The selections in this volume were chosen to be accessible, interesting, informative, and to highlight the attitudes, habits, confidence, persistence and humanity of great problem solvers. The scientists' insatiable desire to learn and to know drove them to ask the right questions and to search for solutions, undaunted by the obstacles that appeared before them. Most of the problems found in this volume took years to solve. There were no answers in the back of the book and more importantly no one, other than nature itself, they could look to for a hint.

The readings are selected from the physical sciences and mathematics[1], arranged in a chronological order that serves to illustrate to some extent the history of thought and discovery from the sixteenth century to present day. The beginning of the scientific revolution finds Copernicus arguing for a sun-centered universe rather than the earth-centered universe as demanded by church dogma. Galileo's discovery

---

[1]   Readings in technology, environmental sciences, and psychology can be found in subsequent volumes of the *Lynchburg College Symposium Readings*.

of the moons of Jupiter further argued against the belief that everything in the universe orbited around the earth. His arguments addressed to the Grand Duchess Christina, along with his careful observations using the newly developed telescope, clearly positioned him in conflict with the church, but clarified the sun as the center of the universe. Newton's pioneering work discovering gravity and in developing calculus showed the relationship between falling objects on earth to the motion of the moon about the earth. Poetry, written in the twentieth century by Siv Cedering, honors the work of the early astronomers reflecting great admiration for their work and contributions to human understanding.

The sixteenth- and seventeenth-century scientists discovered the ordered universe. The universe could now be explained by mathematics which could be used to develop models to show a predictable and understandable world, known as the mechanistic view,[2] which prevails to modern times. Buoyed by technological advancement, including the discovery of electricity, the telescope to see into the heavens, and the microscope to make visible cellular structures to identify many more species not visible to the naked eye, this age of enlightenment fostered the belief that science could explain, even solve, all human problems.

The eighteenth and nineteenth centuries were marked by even more scientific discoveries that set the stage for twentieth-century discoveries in mathematics, physics, chemistry, and biology. Individuals, such as Michael Faraday, who discovered a number of new organic compounds and wrote a manual of practical chemistry, added to the scientific literature. His lectures, designed for young people, on the workings of the candle stands as an important contribution

---

[2]   The mechanistic view of the universe was advanced by the work of Descartes and Francis Bacon. Selected readings from their work, *Discourse on Method* and *The New Organon* can be found in the second edition of *The Lynchburg College Symposium Readings*, Vol. *VII, The Nature of the Universe*. (The University of America Press.)

to the scientific understanding for a larger circle of citizens. James Clerk Maxwell, a man of wide ranging interests, was the first person to propose that light is an electromagnetic phenomenon and that visible light forms only a small part of the whole spectrum of possible electromagnetic radiation. He developed four equations, now called the Maxwell Equations, which completely describe classical electromagnetism. These equations are considered the greatest contributions to nineteenth-century mathematics. His work also set the stage for quantum mechanics and led to modern atomic theory. The mechanized views of Newton began to be more pervasively applied to social and political life. Edwin Abbott, a progressive Anglican clergyman who, in 1884, published a book titled *Flatland* is included in our volume for the first time. This text, a brilliant mathematical description of Euclidean space, is also a pure social satire of the Victorian times in which Abbott lived. The work satirizes the Victorian era's preoccupations with class distinctions, social Darwinism, and resistance to the rights of women and others, all issues which have relevance today.

The science world in the twentieth century was a very different place from the sixteenth, seventeenth, and even the eighteenth or nineteenth centuries. The issue of the earth-centered versus the sun-centered universe was resolved, single cell organisms were commonly understood, and concern for bacteria and germs became the focus of study in the early part of the twentieth century, as did the very important discovery of radiation, and later the development of the atomic bomb. The integrated areas of study in the sciences fractioned into separate disciplines for a good part of the century only to come together again in new fields of study by the end of the twentieth century. Early in the century, Louis Pasteur uncovered the mysteries of rabies and the story of the beginnings of inoculations are recounted in this volume. He also identified the treatments for anthrax, chicken cholera, and contributed to the development of the first vaccines. His work urshered in modern biology and biochemistry. With the process of pasteurization, he gave us the scientific basis for fermentation, wine-making, the brewing of beer, and the careful processing of milk to keep it from spoiling, all pioneering discoveries that enhanced human health and entrepreneurship. Marie and Pierre Curie discovered the phenomenon of radioactivity (a word that she invented) following their discovery of the radioactive elements polonium and radium. Years later, her daughter, Irene, and son-in-law, Frederic, discovered radioisotopes. During World War I, she and Irene developed mobile radiography units for the treatment of wounded soldiers and after the war she devoted herself to developing the medical uses of radiation. Her lifelong focus was the understanding of radioactivity and its potential uses in medicine.

The twentieth century saw chemistry, physics, astronomy, biology, and mathematics become established independent fields of scientific study, often times leading scientists and mathematicians to eschew applied over theoretical studies, each side claiming to address the more important issues. G. H. Hardy, a man of strong, individualistic opinions, argues in *A Mathematicians Apology* that mathematics is beautiful and important for its own sake, and that one should not be concerned about direct applications. His views should be compared to those of Mary Cartwright, a theoretical mathematician who collaborated in the development of chaos theory (see "Mathematics and Thinking Mathematically") who argues that applied applications can be informed and advanced by theoretical mathematics, and that theoretical mathematics can, in turn, be developed from applications. More than twenty years earlier, George Pólya, another theoretical mathematician, in three books on problem solving, provided general heuristics for solving problems of all kinds, not simply mathematical ones. Written for teachers, his focus was on the teaching of problem solving in all areas of human endeavor.

Linus Pauling dominated much of the middle and latter years of the twentieth century with his pioneering work in several fields, especially chemistry. In 1947 he published *General Chemistry*, a book used by generations of undergraduates around the world. As scientific knowledge advanced, more specialized fields of investigation emerged. Pauling is generally credited with establishing molecular biology which serves as the foundation for the emerging field of biotechnology. Following World War II, he became a strong advocate for peace around the world. He devoted his time and energy to establishing world peace and arguing against the further use of the atomic bomb, for him an unfortunate application of scientific know-how. By the end of the century, knowledge gained from in-depth study of many fields within the arts and sciences came together under the umbrella of the cognitive sciences and artificial intelligence. Work by such contemporary thinkers as Douglas Hofstadter is focused on the building of computer models of human thinking. Students will enjoy studying the "MU Puzzle" as an illustration of complex problem solving. The reading by Richard Feynman from his book, *QED: The Strange Theory of Light and Matter*, a series of four lectures that he gave in New Zealand on quantum electrodynamics (QED), explains the parameters of QED, or the interactions between electrons and light.

The readings in this volume are appropriate for, but certainly not limited to, introductory and upper level science courses and the history of the sciences. However the selections were chosen so that no scientific background is required to read them. Thus, they would be appropriate for a much wider variety of courses limited only by the instructor's creativity.

Students and faculty are invited to explore how the discovery of new information informed the creation of the various disciplines over the centuries and to contemplate the many contributions to our understanding of our place in the universe.

Kevin Peterson
Peggy Pittas

NICOLAUS COPERNICUS

1473-1543

Commentariolus
ca. 1514

Nicolaus Copernicus—priest, astronomer, and humanist—founded the heliocentric (sun-centered) theory of the solar system. Copernicus was born in the Polish town of Torun, in what was then West Prussia, of a Polish father and a German mother. After beginning his university education of classics, drawing, mathematics, and perspective at Cracow (in Poland) in 1491, he studied in Italy at Bologna and then at Padua in jurisprudence and medicine. He was awarded his doctorate in Canon Law at Ferrara (Italy). In Rome, he lectured on mathematics and astronomy. He became renowned as a court physician at Heilsburg for 6 years (1506-1512).

In 1497 he received a tenured appointment as canon of the cathedral at Fruenburg in East Prussia. In that same year, he made his first recorded astronomical observations. From his knowledge of ancient Greek and Roman writings, he was satisfied that Ptolemy's geocentric view of the universe was not the only one, and this encouraged him to formulate his own theory.

Living in a Europe fiercely dominated by the Catholic Church, which built its traditions on the Ptolemaic view of the earth as the center of the universe, Copernicus realized that an opposing view would threaten deeply held assumptions, established hierarchies, and cultural practices. He was reluctant to publish his views, but between 1510 and 1514, he summarized his theory in a manuscript, the *Commentariolus*, which he circulated only among his friends. In 1540 he finally consented to the publication of his complete work, *On the Revolutions of the Heavenly Bodies,* which he adroitly dedicated to Pope Paul III. Reportedly, Copernicus received the first copy of the newly published book on the last day of his life, May 24, 1543. The book became a standard reference for problems in astronomical research. In effect, it led the way to the scientific revolution as a whole.

Source:

"Nicolaus Copernicus." 2002. School of Mathematics and Statistics, University of St. Andrews, Scotland. 2002. Retrieved January 17, 2007, from http://www-history.mcs.st-andrews. ac.uk/Biographies/Copernicus.html.

Hagen, J.G. 1908. "Nicolaus Copernicus." In *The Catholic Encyclopedia,* Vol.IV. New York: Appleton. Retrieved July 12, 2007, from http://www.newadvent.org/cathen/04251b.htm

Selection From:

Copernicus [Nicolaus]. *ca.*1514. *The Commentariolus of Copernicus.* In *Three Copernican Treatises.* Trans. Edward Rosen. New York: Dover. 1959. 57-90.

Nicholas Copernicus
Sketch Of His Hypotheses
For The Heavenly Motions

OUR ANCESTORS assumed, I observe, a large number of celestial spheres for this reason especially, to explain the apparent motion of the planets by the principle of regularity. For they thought it altogether absurd that a heavenly body, which is a perfect sphere, should not always move uniformly. They saw that by connecting and combining regular motions in various ways they could make any body appear to move to any position.

Callippus and Eudoxus, who endeavored to solve the problem by the use of concentric spheres, were unable to account for all the planetary movements; they had to explain not merely the apparent revolutions of the planets but also the fact that these bodies appear to us sometimes to mount higher in the heavens, sometimes to descend; and this fact is incompatible with the principle of concentricity. Therefore it seemed better to employ eccentrics and epicycles, a system which most scholars finally accepted.

Yet the planetary theories of Ptolemy and most other astronomers, although consistent with the numerical data, seemed likewise to present no small difficulty. For these theories were not adequate unless certain equants were also conceived; it then appeared that a planet moved with uniform velocity neither on its deferent nor about the center of its epicycle. Hence a system of this sort seemed neither sufficiently absolute nor sufficiently pleasing to the mind.

Having become aware of these defects, I often considered whether there could perhaps be found a more reasonable arrangement of circles, from which every apparent inequality would be derived and in which everything would move uniformly about its proper center, as the rule of absolute motion requires. After I had addressed myself to this very difficult and almost insoluble problem, the suggestion at length came to me how it could be solved with fewer and much simpler constructions than were formerly used, if some assumptions (which are called axioms) were granted me. They follow in this order.

Assumptions

1. There is no one center of all the celestial circles or spheres.
2. The center of the earth is not the center of the universe, but only of gravity and of the lunar sphere.
3. All the spheres revolve about the sun as their mid-point, and therefore the sun is the center of the universe.
4. The ratio of the earth's distance from the sun to the height of the firmament is so much smaller than the ratio of the earth's radius to its distance from the sun that the distance from the earth to the sun is imperceptible in comparison with the height of the firmament.
5. Whatever motion appears in the firmament arises not from any motion of the firmament, but from the earth's motion. The earth together with its circumjacent elements performs a complete rotation on its fixed poles in a daily motion, while the firmament and highest heaven abide unchanged.
6. What appear to us as motions of the sun arise not from its motion but from the motion of the earth and our sphere,

with which we revolve about the sun like any other planet. The earth has, then, more than one motion.

7. The apparent retrograde and direct motion of the planets arises not from their motion but from the earth's. The motion of the earth alone, therefore, suffices to explain so many apparent inequalities in the heavens.

Having set forth these assumptions, I shall endeavor briefly to show how uniformity of the motions can be saved in a systematic way. However, I have thought it well, for the sake of brevity, to omit from this sketch mathematical demonstrations, reserving these for my larger work. But in the explanation of the circles I shall set down here the lengths of the radii; and from these the reader who is not unacquainted with mathematics will readily perceive how closely this arrangement of circles agrees with the numerical data and observations.

Accordingly, let no one suppose that I have gratuitously asserted, with the Pythagoreans, the motion of the earth; strong proof will be found in my exposition of the circles. For the principal arguments by which the natural philosophers attempt to establish the immobility of the earth rest for the most part on the appearances; it is particularly such arguments that collapse here, since I treat the earth's immobility as due to an appearance.

The Order of the Spheres

The celestial spheres are arranged in the following order. The highest is the immovable sphere of the fixed stars, which contains and gives position to all things. Beneath it is Saturn, which Jupiter follows, then Mars. Below Mars is the sphere on which we revolve; then Venus; last is Mercury. The lunar sphere revolves about the center of the earth and moves with the earth like an epicycle. In the same order also, one planet surpasses another in speed of revolution, according as they trace greater or smaller circles. Thus Saturn completes its revolution in thirty years, Jupiter in twelve, Mars in two and one-half, and the earth in one year; Venus in nine months, Mercury in three.

The Apparent Motions of the Sun

The earth has three motions. First, it revolves annually in a great circle about the sun in the order of the signs, always describing equal arcs in equal times; the distance from the center of the circle to the center of the sun is 1/25 of the radius of the circle. The radius is assumed to have a length imperceptible in comparison with the height of the firmament; consequently the sun appears to revolve with this motion, as if the earth lay in the center of the universe. However, this appearance is caused by the motion not of the sun but of the earth, so that, for example, when the earth is in the sign of Capricornus, the sun is seen diametrically opposite in Cancer, and so on. On account of the previously mentioned distance of the sun from the center of the circle, this apparent motion of the sun is not uniform, the maximum inequality being 2 $1/6°$. The line drawn from the sun through the center of the circle is invariably directed toward a point of the firmament about 10° west of the more brilliant of the two bright stars in the head of Gemini; therefore when the earth is opposite this point, and the center of the circle lies between them, the sun is seen at its greatest distance from the earth. In this circle, then, the earth revolves together with whatever else is included within the lunar sphere.

The second motion, which is peculiar to the earth, is the daily rotation on the poles in the order of the signs, that is, from west to east. On account of this rotation the entire universe appears to revolve with enormous speed. Thus does the earth rotate together with its circumjacent waters and encircling atmosphere.

The third is the motion in declination. For the axis of the daily rotation is not parallel to the axis of the great circle, but is inclined to it at an angle that intercepts a portion of a circumference, in our time about 23 1/2°. Therefore, while the center of the earth always remains in the plane of the ecliptic, that is, in the circumference of the great circle, the poles of the earth rotate, both of them describing small circles about centers equidistant from the axis of the great circle. The period of this motion is not quite a year and is nearly equal to the annual revolution on the great circle. But the axis of the great circle is invariably directed toward the points of the firmament which are called the poles of the ecliptic. In like manner the motion in declination, combined with the annual motion in their joint effect upon the poles of the daily rotation, would keep these poles constantly fixed at the same points of the heavens, if the periods of both motions were exactly equal. Now with the long passage of time it has become clear that this inclination of the earth to the firmament changes. Hence it is the common opinion that the firmament has several motions in conformity with a law not yet sufficiently understood. But the motion of the earth can explain all these changes in a less surprising way. I am not concerned to state what the path of the poles is. I am aware that, in lesser matters, a magnetized iron needle always points in the same direction. It has nevertheless seemed a better view to ascribe the changes to a sphere, whose motion governs the movements of the poles. This sphere must doubtless be sublunar.

Equal Motion Should Be Measured Not by the Equinoxes but by the Fixed Stars

Since the equinoxes and the other cardinal points of the universe shift considerably, whoever attempts to derive from them the equal length of the annual revolution necessarily falls into error. Different determinations of this length were made in different ages on the basis of many observations. Hipparchus computed it as 365 1/4 days, and Albategnius the Chaldean as 365^d 5^h 46m, that is, $133/5^m$ or 13 $1/3^m$ less than Ptolemy. Hispalensis increased Albategnius's estimate by the 20th part of an hour, since he determined the tropical year as $365^d\, 5^h\, 49^m$.

Lest these differences should seem to have arisen from errors of observation, let me say that if anyone will study the details carefully, he will find that the discrepancy has always corresponded to the motion of the equinoxes. For when the cardinal points moved 1° in 100 years, as they were found to be moving in the age of Ptolemy, the length of the year was then what Ptolemy stated it to be. When however in the following centuries they moved with greater rapidity, being opposed to lesser motions, the year became shorter; and this decrease corresponded to the increase in precession. For the annual motion was completed in a shorter time on account of the more rapid recurrence of the equinoxes. Therefore the derivation of the equal length of the year from the fixed stars is more accurate. I used Spica Virginis and found that the year has always been 365 days, 6 hours, and about 10 minutes, which is also the estimate of the ancient Egyptians. The same method must be employed also with the other motions of the planets, as is shown by their apsides, by the fixed laws of their motion in the firmament, and by heaven itself with true testimony.

The Moon

The moon seems to me to have four motions in addition to the annual revolution which has been mentioned. For it revolves once a month on its deferent circle about the center of the earth in the order of the signs. The deferent carries the epicycle which is commonly called the epicycle of the first inequality or argument, but which I call the first or greater epicycle. In the upper portion of its circumference this greater epicycle revolves in the direction opposite to that of the deferent, and its period is a little more than a month. Attached to it is a second epicycle. The moon, finally, moving with this second epicycle, completes two revolutions a month in the direction opposite to that of the greater epicycle, so that whenever the center of the greater epicycle crosses the line drawn from the center of the great circle through the center of the earth (I call this line the diameter of the great circle), the moon is nearest to the center of the greater epicycle. This occurs at new and full moon; but contrariwise at the quadratures, midway between new and full moon, the moon is most remote from the center of the greater epicycle. The length of the radius of the greater epicycle is to the radius of the deferent as 1 1/18:10; and to the radius of the smaller epicycle as 4 3/4:1.

By reason of these arrangements the moon appears, at times rapidly, at times slowly, to descend and ascend; and to this first inequality the motion of the smaller epicycle adds two irregularities. For it withdraws the moon from uniform motion on the circumference of the greater epicycle, the maximum inequality being 12 1/4° of a circumference of corresponding size or diameter; and it brings the center of the greater epicycle at times nearer the moon, at times further from it, within the limits of the radius of the smaller epicycle. Therefore, since the moon describes unequal circles about the center of the greater epicycle, the first inequality varies considerably. In conjunctions and oppositions to the sun its greatest value does not exceed 4°56', but in the quadratures it increases to 6°36'. Those who employ an eccentric circle to account for this variation improperly treat the motion on the eccentric as unequal, and, in addition, fall into two manifest errors. For the consequence by mathematical analysis is that when the moon is in quadrature, and at the same time in the lowest part of the epicycle, it should appear nearly four times greater (if the entire disk were luminous) than when new and full, unless its magnitude increases and diminishes in no reasonable way. So too, because the size of the earth is sensible in comparison with its distance from the moon, the lunar parallax should increase very greatly at the quadratures.

But if anyone investigates these matters carefully, he will find that in both respects the quadratures differ very little from new and full moon, and accordingly will readily admit that my explanation is the truer.

With these three motions in longitude, then, the moon passes through the points of its motion in latitude. The axes of the epicycles are parallel to the axis of the deferent, and therefore the moon does not move out of the plane of the deferent. But the axis of the deferent is inclined to the axis of the great circle or ecliptic; hence the moon moves out of the plane of the ecliptic. Its inclination is determined by the size of an angle which subtends 5° of the circumference of a circle. The poles of the deferent revolve at an equal distance from the axis of the ecliptic, in nearly the same manner as was explained regarding declination. But in the present case they move in the reverse order of the signs and much more slowly, the period of the revolution being nineteen years. It is the common opinion that the motion takes place in a higher sphere, to which the poles are attached as they revolve in the manner described. Such a fabric of motions, then, does the moon seem to have.

The Three Superior Planets
Saturn—Jupiter—Mars

Saturn, Jupiter, and Mars have a similar system of motions, since their deferents completely enclose the great circle and revolve in the order of the signs about its center as their common center. Saturn's deferent revolves in 30 years, Jupiter's in 12 years, and that of Mars in 29 months; it is as though the size of the circles delayed the revolutions. For if the radius of the great circle is divided into 25 units, the radius of Mars'[s] deferent will be 38 units, Jupiter's 130 5/12, and Saturn's 230 1/6. By "radius of the deferent" I mean the distance from the center of the deferent to the center of the first epicycle. Each deferent has two epicycles, one of which carries the other, in much the same way as was explained in the case of the moon, but with a different arrangement. For the first epicycle revolves in the direction opposite to that of the deferent, the periods of both being equal. The second epicycle, carrying the planet, revolves in the direction opposite to that of the first with twice the velocity. The result is that whenever the second epicycle is at its greatest or least distance from the center of the deferent, the planet is nearest to the center of the first epicycle; and when the second epicycle is at the midpoints, a quadrant's distance from the two points just mentioned, the planet is most remote from the center of the first epicycle. Through the combination of these motions of the deferent and epicycles, and by reason of the equality of their revolutions, the aforesaid withdrawals and approaches occupy absolutely fixed places in the firmament, and everywhere exhibit unchanging patterns of motion. Consequently the apsides are invariable; for Saturn, near the star which is said to be on the elbow of Sagittarius; for Jupiter, 8° east of the star which is called the end of the tail of Leo; and for Mars, 6 1/2° west of the heart of Leo.

The radius of the great circle was divided above into 25 units. Measured by these units, the sizes of the epicycles are as follows. In Saturn the radius of the first epicycle consists of 19 units, 41 minutes; the radius of the second epicycle, 6 units, 34 minutes. In Jupiter the first epicycle has a radius of 10 units, 6 minutes; the second, 3 units, 22 minutes. In Mars the first epicycle, 5 units, 34 minutes; the second, 1 unit, 51 minutes. Thus the radius of the first epicycle in each case is three times as great as that of the second.

The inequality which the motion of the epicycles imposes upon the motion of the deferent is called the first inequality; it follows, as I have said, unchanging paths everywhere in the firmament. There is a second inequality, on account of which the planet seems from time to time to retrograde, and often to become stationary. This happens by reason of the motion, not of the planet, but of the earth changing its position in the great circle. For since the earth moves more rapidly than the planet, the line of sight directed toward the firmament regresses, and the earth more than neutralizes the motion of the planet. This regression is most notable when the earth is nearest to the planet, that is, when it comes between the sun and the planet at the evening rising of the planet. On the other hand, when the planet is setting in the evening or rising in the morning, the earth makes the observed motion greater than the actual. But when the line of sight is moving in the direction opposite to that of the planets and at an equal rate, the planets appear to be stationary, since the opposed motions neutralize each other; this commonly occurs when the angle at the earth between the sun and the planet is 120°. In all these cases, the lower the deferent on which the planet moves, the greater is the inequality. Hence it is smaller in Saturn than in Jupiter, and again greatest in Mars, in accordance with the ratio of the radius of the great circle to the radii of the deferents. The inequality attains its maximum for each planet when the line of sight to the planet is tangent to the circumference of the great circle. In this manner do these three planets move.

In latitude they have a twofold deviation. While the circumferences of the epicycles remain in a single plane with their deferent, they are inclined to the ecliptic. This inclination is governed by the inclination of their axes, which do not revolve, as in the case of the moon, but are directed always toward the same region of the heavens. Therefore the intersections of the deferent and ecliptic (these points of intersection are called the nodes) occupy eternal places in the firmament. Thus the node where the planet begins its ascent toward the north is, for Saturn, 8 1/2° east of the star which is described as being in the head of the eastern of the two Gemini; for Jupiter, 4° west of the same star; and for Mars, 6 1/2° west of Vergiliae. When the planet is at this point and its diametric opposite, it has no latitude. But the greatest latitude, which occurs at a quadrant's distance from the nodes, is subject to a large inequality. For the inclined axes and circles seem to rest upon the nodes, as though swinging from them. The inclination becomes greatest when the earth is nearest to the planet, that is, at the evening rising of the planet; at that time the inclination of the axis is, for Saturn 2 2/3°, Jupiter 1 2/5°, and Mars 1 5/6. On the other hand, near the time of the evening setting and morning rising, when the earth is at its greatest distance from the planet, the inclination is smaller, for Saturn and Jupiter by 5/12°, and for Mars by 1 2/3°. Thus this inequality is most notable in the greatest latitudes, and it becomes smaller as the planet approaches the node, so that it increases and decreases equally with the latitude.

The motion of the earth in the great circle also causes the observed latitudes to change, its nearness or distance increasing or diminishing the angle of the observed latitude, as mathematical analysis demands. This motion in libration occurs along a straight line, but a motion of this sort can be derived from two circles. These are concentric, and one of them, as it revolves, carries with it the inclined poles of the other. The lower circle revolves in the direction opposite to that of the upper, and with twice the velocity. As it revolves, it carries with it the poles of the circle which serves as deferent to the epicycles. The poles of the deferent are inclined to the poles of the circle halfway above at an angle equal to the inclination of these poles to the poles of the highest circle. So much for Saturn, Jupiter, and Mars and the spheres which enclose the earth.

Venus

There remain for consideration the motions which are included within the great circle, that is, the motions of Venus and Mercury. Venus has a system of circles like the system of the superior planets, but the arrangement of the motions is different. The deferent revolves in nine months, as was said above, and the greater epicycle also revolves in nine months. By their composite motion the smaller epicycle is everywhere brought back to the same path in the firmament, and the higher apse is at the point where I said the sun reverses its course. The period of the smaller epicycle is not equal to that of the deferent and greater epicycle, but has a constant relation to the motion of the great circle. For one revolution of the latter the smaller epicycle completes two. The result is that whenever the earth is in the diameter drawn through the apse, the planet is nearest to the center of the greater epicycle; and it is most remote, when the earth, being in the diameter perpendicular to the diameter through the apse, is at a quadrant's distance from the positions just mentioned. The smaller epicycle of the moon moves in very much the same way with relation to the sun. The ratio of the radius of the great circle to the radius of the deferent of Venus is 25:18; the greater epicycle has a value of 5/4 of a unit, and the smaller 1/4.

Venus seems at times to retrograde, particularly when it is nearest to the earth, like the superior planets, but for the opposite reason. For the regression of the superior planets happens because the motion of the earth is more rapid than theirs, but with Venus, because it is slower; and because the superior planets enclose the great circle, whereas Venus is enclosed within it. Hence Venus is never in opposition to the sun, since the earth cannot come between them, but it moves within fixed distances on either side of the sun. These distances are determined by tangents to the circumference drawn from the center of the earth, and never exceed 48° in our observations. Here ends the treatment of Venus'[s] motion in longitude.

Its latitude also changes for a twofold reason. For the axis of the deferent is inclined at an angle of 2 ½°, the node whence the planet turns north being in the apse. However, the deviation which arises from this inclination, although in itself it is one and the same, appears twofold to us. For when the earth is on the line drawn through the nodes of Venus, the deviations on the one side are seen above, and on the opposite side below; these are called the reflexions. When the earth is at a quadrant's distance from the nodes, the same natural inclinations of the deferent appear, but they are called the declinations. In all the other positions of the earth, both latitudes mingle and are combined, each in turn exceeding the other; by their likeness and difference they are mutually increased and eliminated.

The inclination of the axis is affected by a motion in libration that swings, not on the nodes as in the case of the superior planets, but on certain other movable points. These points perform annual revolutions with reference to the planet. Whenever the earth is opposite the apse of Venus, at that time the amount of the libration attains its maximum for this planet, no matter where the planet may then be on the deferent. As a consequence, if the planet is then in the apse or diametrically opposite to it, it will not completely lack latitude, even though it is then in the nodes. From this point the amount of the libration decreases, until the earth has moved through a quadrant of a circle from the aforesaid position, and, by reason of the likeness of their motions, the point of maximum deviation has moved an equal distance from the planet. Here no trace of the deviation is found. Thereafter the descent of the deviation continues. The initial point drops from north to south, constantly increasing its distance from the planet in accordance with the distance of the earth from the apse. Thereby the planet is brought to the part of the circumference which previously was south. Now, however, by the law of opposition, it becomes north and remains so until the limit of the libration is again reached upon the completion of the circle. Here the deviation becomes equal to the initial deviation and once more attains its maximum. Thus the second semicircle is traversed in the same way as the first. Consequently this latitude, which is usually called the deviation, never becomes a south latitude. In the present instance, also, it seems reasonable that these phenomena should be produced by two concentric circles with oblique axes, as I explained in the case of the superior planets.

Mercury

Of all the orbits in the heavens the most remarkable is that of Mercury, which traverses almost untraceable paths, so that it cannot be easily studied. A further difficulty is the fact that the planet, following a course generally invisible in the rays of the sun, can be observed for a very few days only. Yet Mercury too will be understood, if the problem is attacked with more than ordinary ability.

Mercury, like Venus, has two epicycles which revolve on the deferent. The periods of the greater epicycle and deferent are equal, as in the case of Venus. The apse is located $14\ 1/2°$ east of Spica Virginis. The smaller epicycle revolves with twice the velocity of the earth. But by contrast with Venus, whenever the earth is above the apse or diametrically opposite to it, the planet is most remote from the center of the greater epicycle; and it is nearest, whenever the earth is at a quadrant's distance from the points just mentioned. I have said that the deferent of Mercury revolves in three months, that is, in 88 days. Of the 25 units into which I have divided the radius of the great circle, the radius of the deferent of Mercury contains $9\ 2/5$. The first epicycle contains 1 unit, 41 minutes; the second epicycle is $1/3$ as great, that is, about 34 minutes.

But in the present case this combination of circles is not sufficient, though it is for the other planets. For when the earth passes through the above-mentioned positions with respect to the apse the planet appears to move in a much smaller path than is required by the system of circles described above; and in a much greater path, when the earth is at a quadrant's distance from the positions just mentioned. Since no other inequality in longitude is observed to result from this, it may be reasonably explained by a certain approach of the planet to and withdrawal from the center of the deferent along a straight line. This motion must be produced by two small circles stationed about the center of the greater epicycle, their axes being parallel to the axis of the deferent. The center of the greater epicycle, or of the whole epicyclic structure, lies on the circumference of the small circle that is situated between this center and the outer small circle. The distance from this center to the center of the inner circle is exactly equal to the distance from the latter center to the center of the outer circle. This distance has been found to be 14 ½ minutes of one unit of the 25 by which I have measured the relative sizes of all the circles. The motion of the outer small circle performs two revolutions in a tropical year, while the inner one completes four in the same time with twice the velocity in the opposite direction. By this composite motion the centers of the greater epicycle are carried along a straight line, just as I explained with regard to the librations in latitude. Therefore, in the aforementioned positions of the earth with respect to the apse, the center of the greater epicycle is nearest to the center of the deferent; and it is most remote, when the earth is at a quadrant's distance from these positions. When the earth is at the midpoints, that is, 45° from the points just mentioned, the center of the greater epicycle joins the center of the outer small circle, and both centers coincide. The amount of this withdrawal and approach is 29 minutes of one of the abovementioned units. This, then, is the motion of Mercury in longitude.

Its motion in latitude is exactly like that of Venus, but always in the opposite hemisphere. For where Venus is in north latitude, Mercury is in south. Its deferent is inclined to the ecliptic at an angle of 7°. The deviation, which is always south, never exceeds 3/4°. For the rest, what was said about the latitudes of Venus may be understood here also, to avoid repetition.

Then Mercury runs on seven circles in all; Venus on five; the earth on three, and round it the moon on four; finally Mars, Jupiter, and Saturn on five each. Altogether, therefore, thirty-four circles suffice to explain the entire structure of the universe and the entire ballet of the planets.

GALILEO GALILEI

1564-1642

Galileo Galilei was born in Pisa, the first child of Vincenzo Galilei and Guilia Ammannati. At age ten he was sent to Florence where Jacopo Borghini became his private tutor. Later he enrolled at the Camaldolese Monastery at Vallombrosa near Florence where he became a novice with plans to join the Camaldolese Order but his father was determined that he would be a medical doctor.

Acquiescing to his father's wishes, he enrolled in the medical program at the University of Pisa, but such studies never appealed to Galileo. More interested in mathematics and natural philosophy, he took as many courses as he could in these subjects. His parents insisted that Galileo spend his summers reading for his medical degree, but one of his professors, Ostilio Ricci of Tuscan Court, persuaded them that young Galileo was more suited for the study of mathematics. Galileo never finished his medical studies but immersed himself in mathematics in Florence, Siena, and finally in Rome, where he was a student of Christopher Clavius, a Jesuit priest and an astronomer who helped Pope Gregory XIII create the Gregorian calendar.

Galileo's first appointment was a minor one at the University of Pisa in 1599, but after three years he moved into a more important position for the teaching of mathematics. At the University of Padua, he worked on his theory of motion by using inclined planes and the pendulum. He showed that a projectile follows a parabolic path and he formed equations showing the path of falling objects down an incline. In 1609, he learned about the invention of the spyglass and he used the same concept to develop the telescope (which he called the perspicillum) for the study of the stars. He also learned how to grind glass to see the stars at different distances. Within the next two months, it is said that he made more important discoveries about the heavens that would change the world than had anyone before or since.[1]

Based on his early studies, Galileo published *The Starry Messenger* in 1610. In this work he describes seeing mountains on the moon, reports that the Milky Way is comprised of tiny stars, and says that four moons orbit Jupiter. His book caused a stir in the Venetian Senate, but he left Padua to become the Chief Mathematician at the University of Pisa and Mathematician and Philosopher to the Grand Duke of Tuscany. During a visit to Rome in 1611, he was treated like a celebrity and he was made a fellow of the *Accademia dei Lincei* (Academy of Lynxes)[2] His celebrity was to be short-lived.

Galileo continued studying the heavens and discovered that Venus shared phases similar to the moon's and, therefore, theorized that it must orbit the sun rather than the earth. He was an early believer in the work of Copernicus and by his own studies, he came more and more convinced that Copernicus was correct in his heliocentric view of the planets. Galileo probably could have avoided confrontation with the Catholic Church, which to that time had viewed both his observations and Copernicus's work as eloquent mathematical calculations

[1]   Swerdow, In P Machamer (ed.), The Cambridge companion to Galileo (Cambridge, 1998). Cited in http://www-groups.dcs.st-and.ac.uk/~history/Biographies/Galileo.html, J.J. O'Connor and E.F. Robertson, accessed May 23, 2007.

[2]   According to the Accademia dei Lincei, Galileo was made a member in 1609. His biographies all place the date in 1611.

of the heavily bodies that helped to explain the structure of the universe. But the church held that their findings were hypothetical not *physically real*. However, in 1616, Galileo writes the *Letter to the Grand Duchess Christina* in which he attacks the accepted Aristotelian view of the universe, argues that the Holy Scriptures should not be taken literally, and further claims that mathematics and science can prove the physical world. He further fully supports Copernicus's heliocentric view of the universe.

When this letter became public, Pope Paul V ordered the Sacred Congregation of the Index to conduct an investigation of Copernicus's earlier work. This body condemned his findings and ordered Galileo not to follow Copernican thought or to conduct further research on the subject. Soon after the trial, Galileo's friend, Maffeo Barberini, who had an interest in Galileo's work, became Pope Urban VIII. Galileo erroneously believed that the church would no longer object to his support of Copernican theory. He wrote *Dialogue Concerning the Two Chief Systems of the World—Ptolemaic and Copernican* in 1632. This work was banned by the Inquisition and Galileo was ordered to Rome to stand trial. Because of poor health, he was unable to travel until 1633. He was tried and found guilty of a failure to obey the conditions of the 1616 Inquisition and sentenced to life in prison. He was allowed to spend the remainder of his days under house arrest, first in the home of the archbishop of Siena, and later at his own home near Florence. He was monitored by an officer of the Inquisition.

Nevertheless, he managed to complete another book, *Discourses And Mathematical Demonstrations Concerning The Two New Sciences*, which was smuggled out of Italy into Holland where it was published. This book was a refinement of his first book, *De Motu*, and focused on motion and gravity. He died in 1642, condemned by the Catholic Church for heresy, which lasted until October 31, 1992, when Pope John Paul II announced that the church had made errors in judging Galileo.

Sources:

O'Connor, J. J. and E. F. Robertson, Eds. Retrieved May 23, 2007, from http://www-groups.dcs.st-and.ac.uk/~history/Biographies/Galileo.html.

Wilde, Megan, Ed. Retrieved May 23, 2007, from http://galileo.rice.edu/bio/index.html.

The Starry Messenger
1615

Selection From:

Galilei, Galileo. 1615. *The Starry Messenger.* In *Discoveries and Opinions of Galileo*, Stillman Drake, Trans. New York: Doubleday Anchor Books. 1957. 21-58.

GALILEO GALILEI

**THE
STARRY MESSENGER**
Revealing great, unusual, and remarkable
spectacles, opening these
to the consideration of every man,
and especially of philosophers and
astronomers;
AS OBSERVED BY GALILEO GALILEI
Gentleman of Florence
Professor of Mathematics in the
University of Padua,
**WITH THE AID OF A
SPYGLASS**
lately invented by him,
In the surface of the Moon, in innumerable
Fixed Stars, in Nebulae, and above all
in FOUR PLANETS
swiftly revolving about Jupiter at
differing distances and periods,
and known to no one before the
Author recently perceived them
and decided that they should
be named
THE MEDICEAN STARS

Venice
1610

TO THE MOST SERENE
COSIMO II DE' MEDICI
FOURTH GRAND DUKE OF TUSCANY

Surely a distinguished public service has been rendered by
those who have protected from envy the noble achievements of

men who have excelled in virtue, and have thus preserved from oblivion and neglect those names which deserve immortality. In this way images sculptured in marble or cast in bronze have been handed down to posterity; to this we owe our statues, both pedestrian and equestrian; thus have we those columns and pyramids whose expense (as the poet says)[3] reaches to the stars; finally, thus cities have been built to bear the names of men deemed worthy by posterity of commendation to all the ages. For the nature of the human mind is such that unless it is stimulated by images of things acting upon it from without, all remembrance of them passes easily away.

Looking to things even more stable and enduring, others have entrusted the immortal fame of illustrious men not to marble and metal but to the custody of the Muses and to imperishable literary monuments. But why dwell upon these things as though human wit were satisfied with earthly regions and had not dared advance beyond? For, seeking further, and well understanding that all human monuments ultimately perish through the violence of the elements or by old age, ingenuity has in fact found still more incorruptible monuments over which voracious time and envious age have been unable to assert any rights. Thus turning to the sky, man's wit has inscribed on the familiar and everlasting orbs of most bright stars the names of those whose eminent and godlike deeds have caused them to be accounted worthy of eternity in the company of the stars. And so the fame of Jupiter, of Mars, of Mercury, Hercules, and other heroes by whose names the stars are called, will not fade before the extinction of the stars themselves.

Yet this invention of human ingenuity, noble and admirable as it is, has for many centuries been out of style. Primeval heroes are in possession of those bright abodes, and hold them in their own right. In vain did the piety of Augustus attempt to elect Julius Caesar into their number,

---

[3] Propertius iii, 2, 17.

for when he tried to give the name of "Julian" to a star which appeared in his time (one of those bodies which the Greeks call "comets" and which the Romans likewise named for their hairy appearance), it vanished in a brief time and mocked his too ambitious wish. But we are able, most serene Prince, to read Your Highness in the heavens far more accurately and auspiciously. For scarce have the immortal graces of your spirit begun to shine on earth when in the heavens bright stars appear as tongues to tell and celebrate your exceeding virtues to all time. Behold, then, four stars reserved to bear your famous name; bodies which belong not to the inconspicuous multitude of fixed stars, but to the bright ranks of the planets. Variously moving about most noble Jupiter as children of his own, they complete their orbits with marvelous velocity—at the same time executing with one harmonious accord mighty revolutions every dozen years about the center of the universe; that is, the sun.[4]

Indeed, the Maker of the stars himself has seemed by clear indications to direct that I assign to these new planets Your Highness's famous name in preference to all others. For just as these stars, like children worthy of their sire, never leave the side of Jupiter by any appreciable distance, so (as indeed who does not know?) clemency, kindness of heart, gentleness of manner, splendor of royal blood, nobility in public affairs, and excellency of authority and rule have all fixed their abode and habitation in Your Highness. And who, I ask once more, does not know that all these virtues emanate from the benign star of Jupiter, next after God as the source of all things good? Jupiter; Jupiter, I say, at the instant of Your Highness's birth, having already emerged from the turbid mists of the horizon

---

[4]   This is the first published intimation by Galileo that he accepted the Copernicus system. Tycho had made Jupiter revolve about the sun, but considered the earth to be the center of the universe. It was not until 1613, however, that Galileo unequivocally supported Copernicus in print.

and occupied the midst of the heavens, illuminating the eastern sky from his own royal house, looked out from that exalted throne upon your auspicious birth and poured forth all his splendor and majesty in order that your tender body and your mind (already adorned by God with the most noble ornaments) might imbibe with their first breath that universal influence and power.

But why should I employ mere plausible arguments, when I may prove my conclusion absolutely? It pleased Almighty God that I should instruct Your Highness in mathematics, which I did four years ago at that time of year when it is customary to rest from the most exacting studies. And since clearly it was mine by divine will to serve Your Highness and thus to receive from near at hand the rays of your surpassing clemency and beneficence, what wonder is it that my heart is so inflamed as to think both day and night of little else than how I, who am indeed your subject not only by choice but by birth and lineage, may become known to you as most grateful and most anxious for your glory? And so, most serene Cosimo, having discovered under your patronage these stars unknown to every astronomer before me, I have with good right decided to designate them by the august name of your family. And if I am first to have investigated them, who can justly blame me if I likewise name them, calling them the Medicean Stars, in the hope that this name will bring as much honor to them as the names of other heroes have bestowed on other stars? For, to say nothing of Your Highness's most serene ancestors, whose everlasting glory is testified by the monuments of all history, your virtue alone, most worthy Sire, can confer upon these stars an immortal name. No one can doubt that you will fulfill those expectations, high though they are, which you have aroused by the auspicious beginning of your reign, and will not only meet but far surpass them. Thus when you have conquered your equals you may still vie with yourself, and you and your greatness will become greater every day.

Accept then, most clement Prince, this gentle glory reserved by the stars for you. May you long enjoy those blessings which are sent to you not so much from the stars as from God, their Maker and their Governor.

Your Highness's most devoted servant,
GALILEO GALILEI

PADUA, March 12, 1610

ASTRONOMICAL MESSAGE
**Which contains and explains recent observations
made with the aid of a new spyglass[5]
concerning the surface of the moon,
the Milky Way, nebulous stars, and
innumerable fixed stars,
as well as four planets never before seen, and
now named
THE MEDICEAN STARS**

Great indeed are the things which in this brief treatise I propose for observation and consideration by all students of nature. I say great, because of the excellence of the subject itself, the entirely unexpected and novel character of these things, and finally because of the instrument by means of which they have been revealed to our senses.

Surely it is a great thing to increase the numerous host of fixed stars previously visible to the unaided vision, adding

---

[5] The word "telescope" was not coined until 1611. A detailed account of its origins is given by Edward Rosen in *The Naming of the Telescope* (New York, 1947). In the present translation the modern term has been introduced for the sake of dignity and ease of reading, but only after the passage in which Galileo describes the circumstances which led him to construct the instrument (pp.28-29).

countless more which have never before been seen, exposing these plainly to the eye in numbers ten times exceeding the old and familiar stars.

It is a very beautiful thing, and most gratifying to the sight, to behold the body of the moon, distant from us almost sixty earthly radii,[6] as if it were no farther away than two such measures—so that its diameter appears almost thirty times larger, its surface nearly nine hundred times, and its volume twenty-seven thousand times as large as when viewed with the naked eye. In this way one may learn with all the certainty of sense evidence that the moon is not robed in a smooth and polished surface but is in fact rough and uneven, covered everywhere, just like the earth's surface, with huge prominences, deep valleys, and chasms.

Again, it seems to me a matter of no small importance to have ended the dispute about the Milky Way by making its nature manifest to the very senses as well as to the intellect. Similarly it will be a pleasant and elegant thing to demonstrate that the nature of those stars which astronomers have previously called "nebulous" is far different from what has been believed hitherto. But what surpasses all wonders by far, and what particularly moves us to seek the attention of all astronomers and philosophers, is the discovery of four wandering stars not known or observed by any man before us. Like Venus and Mercury, which have their own periods about the sun, these have theirs about a certain star that

---

[6]   The original text reads "diameters" here and in another place. That is error was Galileo's and not the printer's has been convincingly shown by Edward Rosen (*Isis*, 1954, pp. 344ff.) The slip was a curious one, as astronomers of all schools had long agreed that the maximum distance of the moon was approximately sixty terrestrial radii. Still more curious is the fact that neither Kepler nor any other correspondent appears to have called Galileo's attention to this error; not even a friend who ventured to criticize the calculations in this very passage.

is conspicuous among those already known, which they sometimes precede and sometimes follow, without ever departing from it beyond certain limits. All these facts were discovered and observed by me not many days ago with the aid of a spyglass which I devised, after first being illuminated by divine grace. Perhaps other things, still more remarkable, will in time be discovered by me or by other observers with the aid of such an instrument, the form and construction of which I shall first briefly explain, as well as the occasion of its having been devised. Afterwards I shall relate the story of the observations I have made.

About ten months ago a report reached my ears that a certain Fleming[7] had constructed a spyglass by means of which visible objects, though very distant from the eye of the observer, were distinctly seen as if nearby. Of this truly remarkable effect several experiences were related, to which some persons gave credence while others denied them. A few days later the report was confirmed to me in a letter from a noble Frenchman at Paris, Jacques Badovere,[8] which caused me to apply myself wholeheartedly to inquire into the means by which I might arrive at the invention of a similar instrument. This I did shortly afterwards, my basis being the theory of refraction. First I prepared a tube of lead, at the ends of which I fitted two glass lenses, both plane on one side while on the other side one was spherically convex and the other concave. Then placing my eye near the concave lens I perceived objects satisfactorily large and near, for they appeared three times closer and nine times larger than when seen with the naked

---

[7]   Credit for the original invention is generally assigned to Hans Lipperhey, a lens grinder in Holland who chanced upon this property of combined lenses and applied for a patent on it in 1608.

[8]   Badovere studied in Italy toward the close of the sixteenth century and is said to have been a pupil of Galileo's about 1598. When he wrote concerning the new instrument in 1609 he was in the French diplomatic service at Paris, where he died in 1620.

eye alone. Next I constructed another one, more accurate, which represented objects as enlarged more than sixty times. Finally, sparing neither labor nor expense, I succeeded in constructing for myself so excellent an instrument that objects seen by means of it appeared nearly one thousand times larger and over thirty times closer than when regarded with our natural vision.

It would be superfluous to enumerate the number and importance of the advantages of such an instrument at sea as well as on land. But forsaking terrestrial observations, I turned to celestial ones, and first I saw the moon from as near at hand as if it were scarcely two terrestrial radii away. After that I observed often with wondering delight both the planets and the fixed stars, and since I saw these latter to be very crowded, I began to seek (and eventually found) a method by which I might measure their distances apart. Here it is appropriate to convey certain cautions to all who intend to undertake observations of this sort, for in the first place it is necessary to prepare quite a perfect telescope, which will show all objects bright, distinct, and free from any haziness, while magnifying them at least four hundred times and thus showing them twenty times closer. Unless the instrument is of this kind it will be vain to attempt to observe all the things which I have seen in the heavens, and which will presently be set forth. Now in order to determine without much trouble the magnifying power of an instrument, trace on paper the contour of two circles or two squares of which one is four hundred times as large as the other, as it will be when the diameter of one is twenty times that of the other. Then, with both these figures attached to the same wall, observe them simultaneously from a distance, looking at the smaller one through the telescope and at the larger one with the other eye unaided. This may be done without inconvenience while holding both eyes open at the same time; the two figures will appear to be of the same size if the instrument magnifies objects in the desired proportion.

Such an instrument having been prepared, we seek a method of measuring distances apart. This we shall accomplish by the following contrivance.

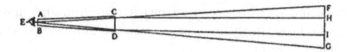

Let ABCD be the tube and E be the eye of the observer. Then if there were no lenses in the tube, the rays would reach the object FG along the straight lines ECF and EDG. But when the lenses have been inserted, the rays go along the refracted lines ECH and EDI; thus they are brought closer together, and those which were previously directed freely to the object FG now include only the portion of it HI. The ratio of the distance EH to the line HI then being found, one may by means of a table of sines determine the size of the angle formed at the eye by the object HI, which we shall find to be but a few minutes of arc. Now, if to the lens CD we fit thin plates, some pierced with larger and some with smaller apertures, putting now one plate and now another over the lens as required, we may form at pleasure different angles subtending more or fewer minutes of arc, and by this means we may easily measure the intervals between stars which are but a few minutes apart, with no greater error than one or two minutes. And for the present let it suffice that we have touched lightly on these matters and scarcely more than mentioned them, as on some other occasion we shall explain the entire theory of this instrument.

Now let us review the observations made during the past two months, once more inviting the attention of all who are eager for true philosophy to the first steps of such important contemplations. Let us speak first of that surface of the moon which faces us. For greater clarity I distinguish two parts of this surface, a lighter and a darker; the lighter part seems to surround and to pervade the whole hemisphere, while the darker part discolors the moon's surface like a kind of cloud, and makes it appear covered with spots. Now those spots which are fairly dark and rather large are plain to everyone and have been seen throughout the ages; these I shall call the "large" or "ancient" spots, distinguishing them from others that are smaller in size but so numerous as to occur all over the lunar surface, and especially the lighter part. The latter spots have never been seen by anyone before me. From observations of these spots repeated many times I have been led to the opinion and conviction that the surface of the moon is not smooth, uniform, and precisely spherical as a great number of philosophers believe it (and the other heavenly bodies) to be, but is uneven, rough, and full of cavities and prominences, being not unlike the face of the earth, relieved by chains of mountains and deep valleys. The things I have seen by which I was enabled to draw this conclusion are as follows.

On the fourth or fifth day after the new moon, when the moon is seen with brilliant horns, the boundary which divides the dark part from the light does not extend uniformly in an oval line as would happen on a perfectly spherical solid, but traces out an uneven, rough, and very wavy line as shown in the figure below. Indeed, many luminous excrescences extend

beyond the boundary to the darker portion, while on the other hand some dark patches invade the illuminated part. Moreover a great quantity of small blackish spots, entirely separated from the dark region, are scattered almost all over the area illuminated by the sun with the exception only of that part which is occupied by the large and ancient spots. Let us note, however, that the said small spots always agree in having their blackened parts directed toward the sun, while on the side opposite the sun they are crowned with bright contours, like shining summits. There is a similar sight on earth about sunrise, when we behold the valleys not yet flooded with light though the mountains surrounding them are already ablaze with glowing splendor on the side opposite the sun. And just as the shadows in the hollows on earth diminish in size as the sun rises higher, so these spots on the moon lose their blackness as the illuminated region grows larger and larger.

Again, not only are the boundaries of shadow and light in the moon seen to be uneven and wavy, but still more astonishingly many bright points appear within the darkened portion of the moon, completely divided and separated from the illuminated part and at a considerable distance from it. After a time these gradually increase in size and brightness, and an hour or two later they become joined with the rest of the lighted part which has now increased in size. Meanwhile more and more peaks shoot up as if sprouting now here, now there, lighting up within the shadowed portion; these become larger, and finally they too are united with that same luminous surface which extends ever further. An illustration of this is to be seen in the figure above. And on the earth, before the rising of the sun, are not the highest peaks of the mountains illuminated by the sun's rays while the plains remain in shadow? Does not the light go on spreading while the larger central parts of those mountains are becoming illuminated? And when the sun has finally risen, does not the illumination of plains and hills finally become one? But on the moon the variety of elevations and depressions appears to surpass in every way the roughness of the terrestrial surface, as we shall demonstrate further on.

At present I cannot pass over in silence something worthy of consideration which I observed when the moon was approaching first quarter, as shown in the previous figure. Into the luminous part there extended a great dark gulf in the neighborhood of the lower cusp. When I had observed it for a long time and had seen it completely dark, a bright peak began to emerge, a little below its center, after about two hours. Gradually growing, this presented itself in a triangular shape, remaining completely detached and separated from the lighted surface. Around it three other small points soon began to shine, and finally, when the moon was about to set, this triangular shape (which had meanwhile become more widely extended) joined with the rest of the illuminated region and suddenly burst into the gulf of shadow like a vast promontory of light, surrounded still by the three bright peaks already mentioned. Beyond the ends of the cusps, both above and below, certain bright points emerged which were quite detached from the remaining lighted part, as may be seen depicted in the same figure. There were also a great number of dark spots in both the horns, especially in the lower one; those nearest the boundary of light and shadow appeared larger and darker, while those more distant from the boundary were not so dark and distinct. But in all cases, as we have mentioned earlier, the blackish portion of each spot is turned toward the source of the sun's radiance, while a bright rim surrounds the spot on the side away from the sun in the direction of the shadowy region of the moon. This part of the moon's surface, where it is spotted as the tail of a peacock is sprinkled with azure eyes, resembles those glass vases which have been plunged while still hot into cold water and have thus acquired a crackled and wavy surface, from which they receive their common name of "ice-cups."

As to the large lunar spots, these are not seen to be broken in the above manner and full of cavities and prominences; rather, they are even and uniform, and brighter patches crop up only here and there. Hence if anyone wished to revive the old Pythagorean[9] opinion that the moon is like another earth, its brighter part might very fitly represent the surface of the land and its darker region that of the water. I have never doubted that if our globe were seen from afar when flooded with sunlight, the land regions would appear brighter and the watery regions darker.[10]

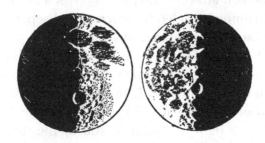

The large spots in the moon are also seen to be less elevated than the brighter tracts, for whether the moon is waxing or waning, there are always seen, here and there along its boundary of light and shadow, certain ridges of brighter hue around the large spots (and we have attended to this in preparing the diagrams); the edges of these spots are not only lower, but also more uniform, being uninterrupted by peaks or ruggedness.

---

[9]   Pythagoras was a mathematician and philosopher of the sixth century B.C., a semilegendary figure whose followers were credited at Galileo's time with having anticipated the Copernican system. This tradition was based upon a misunderstanding. The Pythagoreans made the earth revolve about a "central fire" whose light and heat were reflected to the earth by the sun.

[10]   Leonardo da Vinci had previously suggested that the dark and light regions of the moon were bodies of land and water, though Galileo probably did not know this. Da Vinci, however, had mistakenly supposed that the water would appear brighter than the land.

Near the large spots the brighter part stands out particularly in such a way that before first quarter and toward last quarter, in the vicinity of a certain spot in the upper (or northern) region of the moon, some vast prominences arise both above and below as shown in the figures reproduced below. Before last quarter this same spot is seen to be walled about with certain blacker contours which, like the loftiest mountaintops, appear darker on the side away from the sun and brighter on that which faces the sun. (This is the opposite of what happens in the cavities, for there the part away from the sun appears brilliant, while that which is turned toward the sun is dark and in shadow.) After a time, when the lighted portion of the moon's surface has diminished in size and when all (or nearly all) the said spot is covered with shadow, the brighter ridges of the mountains gradually emerge from the shade. This double aspect of the spot is illustrated in the ensuing figures.

There is another thing which I must not omit, for I beheld it not without a certain wonder; this is that almost in the center of the moon there is a cavity larger than all the rest, and perfectly round in shape. I have observed it near both first and last quarters, and have tried to represent it as correctly as possible in the second of the above figures. As to light and shade, it offers the same appearance as would a region like Bohemia[11] if that were enclosed on all sides by very lofty mountains arranged exactly in a circle. Indeed, this area on the moon is surrounded by such enormous peaks that the bounding edge adjacent to the dark portion of the moon is seen to be bathed

---

[11] This casual comparison between a part of the moon and a specific region on earth was later the basis of much trouble for Galileo; see the letter of G. Ciampoli, p. 158. Even in antiquity the idea that the moon (or any other heavenly body) was of the same nature of the earth had been dangerous to hold. The Athenians banished the philosopher Anaxagoras for teaching such notions, and charged Socrates with blasphemy for repeating them.

in sunlight before the boundary of light and shadow reaches halfway across the same space. As in other spots, its shaded portion faces the sun while its lighted part is toward the dark side of the moon; and for a third time I draw attention to this as a very cogent proof of the ruggedness and unevenness

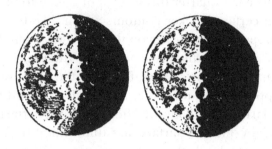

that pervades all the bright region of the moon. Of these spots, moreover, those are always darkest which touch the boundary line between light and shadow, while those farther off appear both smaller and less dark, so that when the moon ultimately becomes full (at opposition[12] to the Sun), the shade of the cavities is distinguished from the light of the places in relief by a subdued and very tenuous separation.

The things we have reviewed are to be seen in the brighter region of the moon. In the large spots, no such contrast of depressions and prominences is perceived as that which we are compelled to recognize in the brighter parts by the changes of aspect that occur under varying illumination by the sun's rays throughout the multiplicity of positions from which the latter reach the moon. In the large spots there exist some holes rather darker than the rest, as we have shown in the illustrations. Yet these present always the same appearance, and their darkness

---

[12]   Opposition of the sun and moon occurs when they are in line with the earth between them (full moon, or lunar eclipse); conjunction, when they are in line on the same side of the earth (new moon, or eclipse of the sun).

is neither intensified nor diminished, although with some minute difference they appear sometimes a little more shaded and sometimes a little lighter according as the rays of the sun fall on them more or less obliquely. Moreover, they join with the neighboring regions of the spots in a gentle linkage, the boundaries mixing and mingling. It is quite different with the spots which occupy the brighter surface of the moon; these, like precipitous crags having rough and jagged peaks, stand out starkly in sharp contrasts of light and shade. And inside the large spots there are observed certain other zones that are brighter, some of them very bright indeed. Still, both these and the darker parts present always the same appearance; there is no change either of shape or of light and shadow; hence one may affirm beyond any doubt that they owe their appearance to some real dissimilarity of parts. They cannot be attributed merely to irregularity of shape, wherein shadows move in consequence of varied illuminations from the sun, as indeed is the case with the other, smaller, spots which occupy the brighter part of the moon and which change, grow, shrink, or disappear from one day to the next, as owing their origin only to shadows of prominences.

But here I foresee that many persons will be assailed by uncertainty and drawn into a grave difficulty, feeling constrained to doubt a conclusion already explained and confirmed by many phenomena. If that part of the lunar surface which reflects sunlight more brightly is full of chasms (that is, of countless prominences and hollows), why is it that the western edge of the waxing moon, the eastern edge of the waning moon, and the entire periphery of the full moon are not seen to be uneven, rough, and wavy? On the contrary they look as precisely round as if they were drawn with a compass; and yet the whole periphery consists of that brighter lunar substance which we have declared to be filled with heights and chasms. In fact not a single one of the great spots extends to the extreme periphery of the moon, but all are grouped together at a distance from the edge.

Now let me explain the twofold reason for this troublesome fact, and in turn give a double solution to the difficulty. In the first place, if the protuberances and cavities in the lunar body existed only along the extreme edge of the circular periphery bounding the visible hemisphere, the moon might (indeed, would necessarily) look to us almost like a toothed wheel, terminated by a warty or wavy edge. Imagine, however, that there is not a single series of prominences arranged only along the very circumference, but a great many ranges of mountains together with their valleys and canyons disposed in ranks near the edge of the moon, and not only in the hemisphere visible to us but everywhere near the boundary line of the two hemispheres. Then an eye viewing them from afar will not be able to detect the separation of prominences by cavities, because the intervals between the mountains located in a given circle or a given chain will be hidden by the interposition of other heights situated in yet other ranges. This will be especially true if the eye of the observer is placed in the same straight line with the summits of these elevations. Thus on earth the summits of several mountains close together appear to be situated in one plane if the spectator is a long way off and is placed at an equal elevation. Similarly in a rough sea the tops of the waves seem to lie in one plane, though between one high crest and another there are many gulfs and chasms of such depth as not only to hide the hulls but even the bulwarks, masts, and rigging of stately ships. Now since there are many chains of mountains and chasms on the moon in addition to those around its periphery, and since the eye, regarding these from a great distance, lies nearly in the plane of their summits, no one need wonder that they appear as arranged in a regular and unbroken line.

To the above explanation another may be added; namely, that there exists around the body of the moon, just as around the earth, a globe of some substance denser than the rest of the aether.[13] This may serve to receive and reflect the sun's radiations without being sufficiently opaque to prevent our seeing through it, especially when it is not illuminated. Such a globe, lighted by the sun's rays, makes the body of the moon appear larger than it really is, and if it were thicker it would be able to prevent our seeing the actual body of the moon. And it actually is thicker near the circumference of the moon; I do not mean in an absolute sense, but relatively to the rays of our vision, which cut it obliquely there. Thus it may obstruct our vision, especially when it is lighted, and cloak the lunar periphery that is exposed to the sun. This may be more clearly understood from the figure below, in which the body of the moon, ABC, is surrounded by the vaporous globe DEG.

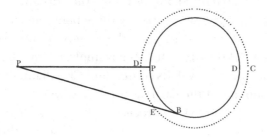

The eyesight from F reaches the moon in the central region, at A for example, through a lesser thickness of the vapors DA, while toward the extreme edges a deeper stratum of vapors, EB, limits and shuts out our sight. One indication of this is that the illuminated portion of the moon appears to be larger in circumference than the rest of the orb, which lies in shadow. And perhaps this same cause will appeal to some as

---

13   The aether, or ever-moving," was the special substance of which the sky and all the heavenly bodies were supposed to be made, a substance essentially different from all the earthly "elements." In later years Galileo abandoned his suggestion here that the moon has a vaporous atmosphere.

reasonably explaining why the larger spots on the moon are nowhere seen to reach the very edge probable though it is that some should occur there. Possibly they are invisible by being hidden under a thicker and more luminous mass of vapors.

That the lighter surface of the moon is everywhere dotted with protuberances and gaps has, I think, been made sufficiently clear from the appearances already explained. It remains for me to speak of their dimensions, and to show that the earth's irregularities are far less than those of the moon. I mean that they are absolutely less, and not merely in relation to the sizes of the respective globes. This is plainly demonstrated as follows.

I had often observed, in various situations of the moon with respect to the sun, that some summits within the shadowy portion appeared lighted, though lying some distance from the boundary of the light. By comparing this separation to the whole diameter of the moon, I found that it sometimes exceeded one-twentieth of the diameter. Accordingly, let CAF be a great circle of the lunar body, E its center, and CF a diameter, which is to the diameter of the earth as two is to seven.

Since according to very precise observations the diameter of the earth is seven thousand miles, CF will be two thousand, CE one thousand, and one-twentieth of CF will be one hundred miles. Now let CF be the diameter of the great circle which divides the light part of the moon from the dark part (for because of the very great distance of the sun from the moon, this does not differ appreciably from a great Circle), and let A be distant from C by one-twentieth of this.

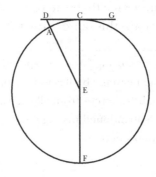

Draw the radius EA, which, when produced, cuts the tangent line GCD (representing the illuminating ray) in the point D. Then the arc CA, or rather the straight line CD, will consist of one hundred units whereof CE contains one thousand, and the sum of the squares of DC and CE will be 1,010,000. This is equal to the square of DE; hence ED will exceed 1,004, and AD will be more than four of those units of which CE contains one thousand. Therefore the altitude AD on the moon, which represents a summit reaching up to the solar ray GCD and standing at the distance CD from C, exceeds four miles. But on the earth we have no mountains which reach to a perpendicular height of even one mile.[14] Hence it is quite clear that the prominences on the moon are loftier than those on the earth.

Here I wish to assign the cause of another lunar phenomenon well worthy of notice. I observed this not just recently, but many years ago, and pointed it out to some of my friends and pupils, explaining it to them and giving its true cause. Yet since it is rendered more evident and easier to observe with the aid of the telescope, I think it not unsuitable for introduction in this place, especially as it shows more clearly the connection between the moon and the earth.

When the moon is not far from the sun, just before or after new moon, its globe offers itself to view not only on the side where it is adorned with shining horns, but a certain faint light is also seen to mark out the periphery of the dark part which faces away from the sun, separating this from the darker background of the aether. Now if we examine the matter more closely, we shall see that not only does the extreme limb of the

---

[14]   Galileo's estimate of four miles for the height of some lunar mountains was a very good one. His remark about the maximum height of mountains on the earth was, however, quite mistaken. An English propagandist for his views, John Wilkins, took plains to correct this error in his anonymous *Discovery of a New World . . . in the Moon* (London, 1638), Prop. ix.

shaded side glow with this uncertain light, but the entire face of the moon (including the side which does not receive the glare of the sun) is whitened by a not inconsiderable gleam. At first glance only a thin luminous circumference appears, contrasting with the darker sky coterminous with it; the rest of the surface appears darker from its contact with the shining horns which distract our vision. But if we place ourselves so as to interpose a roof or chimney or some other object at a considerable distance from the eye, the shining horns may be hidden while the rest of the lunar globe remains exposed to view. It is then found that this region of the moon, though deprived of sunlight, also shines not a little. The effect is heightened if the gloom of night has already deepened through departure of the sun, for in a darker field a given light appears brighter.

Moreover, it is found that this secondary light of the moon (so to speak) is greater according as the moon is closer to the sun. It diminishes more and more as the moon recedes from that body until, after the first quarter and before the last, it is seen very weakly and uncertainly even when observed in the darkest sky. But when the moon is within sixty degrees of the sun it shines remarkably, even in twilight; so brightly indeed that with the aid of a good telescope one may distinguish the large spots. This remarkable gleam has afforded no small perplexity to philosophers, and in order to assign a cause for it some have offered one idea and some another. Some would say it is an inherent and natural light of the moon's own; others, that it is imparted by Venus; others yet, by all the stars together; and still others derive it from the sun, whose rays they would have permeate the thick solidity of the moon. But statements of this sort are refuted and their falsity evinced with little difficulty. For if this kind of light were the moon's own, or were contributed by the stars, the moon would retain it and would display it particularly during eclipses, when it is left in an unusually dark sky. This is contradicted by experience, for the brightness which is seen on the moon during eclipses is much fainter and is ruddy, almost copper-colored, while this is brighter and whitish. Moreover the other light is variable and movable, for it covers the face of the moon in such a way that the place near the edge of the earth's shadow is always seen to be brighter than the rest of the moon; this undoubtedly results from contact of the tangent solar rays with some denser zone which girds the moon about.[15] By this contact a sort of twilight is diffused over the neighboring regions of the moon, just as on earth a sort of crepuscular light is spread both morning and evening; but with this I shall deal more fully in my book on the system of the world.[16]

To assert that the moon's secondary light is imparted by Venus is so childish as to deserve no reply. Who is so ignorant as not to understand that from new moon to a separation of sixty degrees between moon and sun, no part of the moon which is averted from the sun can possibly be seen from Venus? And it is likewise unthinkable that this light should depend upon the sun's rays penetrating the thick solid mass of the moon, for then this light would never dwindle, inasmuch as one hemisphere of the moon is always illuminated except during lunar eclipses. And the light does diminish as the moon approaches first quarter, becoming completely obscured after that is passed.

Now since the secondary light does not inherently belong to the moon, and is not received from any star or from the

---

[15] Kepler had correctly accounted for the existence of this light and its ruddy color. It is caused by refraction of sunlight in the earth's atmosphere, and does not require a lunar atmosphere as supposed by Galileo.

[16] The book thus promised was destined not to appear for more than two decades. Events which will presently be recounted prevented its publication for many years, and then it had to be modified to present the arguments for both the Ptolemaic and Copernican systems instead of just the latter as Galileo here planned. Even then it was suppressed, and the author was condemned to life imprisonment.

sun, and since in the whole universe there is no other body left but the earth, what must we conclude? What is to be proposed? Surely we must assert that the lunar body (or any other dark and sunless orb) is illuminated by the earth. Yet what is there so remarkable about this? The earth, in fair and grateful exchange, pays back to the moon an illumination similar to that which it receives from her throughout nearly all the darkest gloom of night.

Let us explain this matter more fully. At conjunction the moon occupies a position between the sun and the earth; it is then illuminated by the sun's rays on the side which is turned away from the earth. The other hemisphere, which faces the earth, is covered with darkness; hence the moon does not illuminate the surface of the earth at all. Next, departing gradually from the sun, the moon comes to be lighted partly upon the side it turns toward us, and its whitish horns, still very thin, illuminate the earth with a faint light. The sun's illumination of the moon increasing now as the moon approaches first quarter, a refection of that light to the earth also increases. Soon the splendor on the moon extends into a semicircle, and our nights grow brighter; at length the entire visible face of the moon is irradiated by the sun's resplendent rays, and at full moon the whole surface of the earth shines in a flood of moonlight. Now the moon, waning, sends us her beams more weakly, and the earth is less strongly lighted; at length the moon returns to conjunction with the sun, and black night covers the earth.

In this monthly period, then, the moonlight gives us alternations of brighter and fainter illumination; and the benefit is repaid by the earth in equal measure. For while the moon is between us and the sun (at new moon), there lies before it the entire surface of that hemisphere of the earth which is exposed to the sun and illuminated by vivid rays. The moon receives the light which this reflects, and thus the nearer hemisphere of the moon—that is, the one deprived of sunlight—appears by virtue of this illumination to be not a little luminous. When the moon is ninety degrees away from the sun it sees but half the earth illuminated (the western half), for the other (the eastern half) is enveloped in night. Hence the moon itself is illuminated less brightly from the earth, and as a result its secondary light appears fainter to us. When the moon is in opposition to the sun, it faces a hemisphere of the earth that is steeped in the gloom of night, and if this position occurs in the plane of the ecliptic the moon will receive no light at all, being deprived of both the solar and the terrestrial rays. In its various other positions with respect to the earth and sun, the moon receives more or less light according as it faces a greater or smaller portion of the illuminated hemisphere of the earth. And between these two globes a relation is maintained such that whenever the earth is most brightly lighted by the moon, the moon is least lighted by the earth, and vice versa.

Let these few remarks suffice us here concerning this matter, which will be more fully treated in our *System of the world*. In that book, by a multitude of arguments and experiences, the solar reflection from the earth will be shown to be quite real—against those who argue that the earth must be excluded from the dancing whirl of stars for the specific reason that it is devoid of motion and of light. We shall prove the earth to be a wandering body surpassing the moon in splendor, and not the sink of all dull refuse of the universe; this we shall support by an infinitude of arguments drawn from nature.

Thus far we have spoken of our observations concerning the body of the moon. Let us now set forth briefly what has thus far been observed regarding the fixed stars. And first of all, the following fact deserves consideration: The stars, whether fixed or wandering,[17] appear not to be enlarged by the telescope

---

[17]   That is, planets. Among these bodies Galileo counted his newly discovered satellites of Jupiter. The term "satellites" was introduced somewhat later by Kepler.

in the same proportion as that in which it magnifies other objects, and even the moon itself. In the stars this enlargement seems to be so much less that a telescope which is sufficiently powerful to magnify other objects a hundredfold is scarcely able to enlarge the stars four or five times. The reason for are this is as follows.

When stars are viewed by means of unaided natural vision, they present themselves to us not as of their simple (and, so to speak, their physical) size, but as irradiated by a certain fulgor and as fringed with sparkling rays, especially when the night is far advanced. From this they appear larger than they would if stripped of those adventitious hairs of light, for the angle at the eye is determined not by the primary body of the star but by the brightness which extends so widely about it. This appears quite clearly from the fact that when stars first emerge from twilight at sunset they look very small, even if they are of the first magnitude; Venus itself, when visible in broad daylight, is so small as scarcely to appear equal to a star of the sixth magnitude. Things fall out differently with other objects, and even with the moon itself; these, whether seen in daylight or the deepest night, appear always of the same bulk. Therefore the stars are seen crowned among shadows, while daylight is able to remove their headgear; and not daylight alone, but any thin cloud that interposes itself between a star and the eye of the observer. The same effect is produced by black veils or colored glasses, through the interposition of which obstacles the stars abandoned by their surrounding brilliance. A telescope similarly accomplishes the same result. It removes from the stars their adventitious and accidental rays, and then it enlarges their simple globes (if indeed the stars are naturally globular) so that they seem to be magnified in a lesser ratio than other objects. In fact a star of the fifth or sixth magnitude when seen through a telescope presents itself as one of the first magnitude.

Deserving of notice also is the difference between the appearances of the planets and of the fixed stars.[18] The planets show their globes perfectly round and definitely bounded, looking like little moons, spherical and flooded all over with light; the fixed stars are never seen to be bounded by a circular periphery, but have rather the aspect of blazes whose rays vibrate about them and scintillate a great deal. Viewed with a telescope they appear of a shape similar to that which they present to the naked eye, but sufficiently enlarged so that a star of the fifth or sixth magnitude seems to equal the Dog Star, largest of all the fixed stars. Now, in addition to stars of the sixth magnitude, a host of other stars are perceived through the telescope which escape the naked eye; these are so numerous as almost to surpass belief. One may, in fact, see more of them than all the stars included among the first six magnitudes. The largest of these, which we may call stars of the seventh magnitude, or the first magnitude of invisible stars, appear through the telescope as larger and brighter than stars of the second magnitude when the latter are viewed with the naked eye. In order to give one or two proofs of their almost inconceivable number, I have adjoined pictures of two constellations. With these as samples, you may judge of all the others.

In the first I had intended to depict the entire constellation of Orion, but I was overwhelmed by the vast quantity of stars and by limitations of time, so I have deferred this to another occasion. There are more than five hundred new stars distributed among

---

[18] Fixed stars are so distant that their light reaches the earth as from dimensionless points. Hence their images are not enlarged by even the best telescopes, which serve only to gather more of their light and in that way increase their visibility. Galileo was never entirely clear about this distinction. Nevertheless, by applying his knowledge of the effects described here, he greatly reduced the prevailing overestimation of visual dimensions of stars and planets.

the old ones within limits of one or two degrees of arc. Hence to the three stars in the Belt of Orion and the six in the Sword which were previously known, I have added eighty adjacent stars discovered recently, preserving the intervals between them as exactly as I could. To distinguish the known or ancient stars, I have depicted them larger and have outlined them doubly; the other (invisible) stars I have drawn smaller and without the extra line. I have also preserved differences of magnitude as well as possible.

The Belt and Sword of Orion

In the second example I have depicted the six stars of Taurus known as the Pleiades (I say six, inasmuch as the seventh is hardly ever visible) which lie within very narrow limits in the sky. Near them are more than forty others, invisible, no one of which is much more than half a degree away from the original six. I have shown thirty-six of these in the diagram; as in the case of Orion I have preserved their intervals and magnitudes, as well as the distinction between old stars and new.

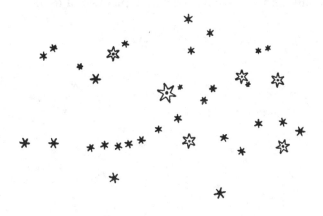

The Pleiades

Third, I have observed the nature and the material of the Milky Way. With the aid of the telescope this has been scrutinized so directly and with such ocular certainty that all the disputes which have vexed philosophers through so many ages have been resolved, and we are at last freed from wordy debates about it. The galaxy is, in fact, nothing but a congeries of innumerable stars grouped together in clusters. Upon whatever part of it the telescope is directed, a vast crowd of stars is immediately presented to view. Many of them are rather large and quite bright, while the number of smaller ones is quite beyond calculation.

But it is not only in the Milky Way that whitish clouds are seen; several patches of similar aspect shine with faint light here and there throughout the aether, and if the telescope is turned upon any of these it confronts us with a tight mass of stars. And what is even more remarkable, the stars which have been called "nebulous" by every astronomer up to this time turn out to be groups of very small stars arranged in a wonderful manner. Although each star separately escapes our sight on account of its smallness or the immense distance from us, the mingling of their rays gives rise to that gleam which was formerly believed to be some denser part of the aether that was capable of reflecting rays from stars or from the sun. I have observed some of these constellations and have decided to depict two of them.

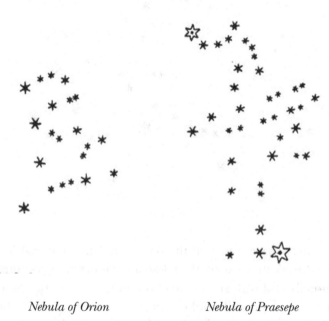

Nebula of Orion   *Nebula of Praesepe*

In the first you have the nebula called the Head of Orion, in which I have counted twenty-one stars. The second contains the nebula called Praesepe,[19] which is not a single star but a mass of more than forty starlets. Of these I have shown thirty-six, in addition to the Aselli, arranged in the order shown.

---

[19]   Praesepe, "the Manager," is a small whitish cluster of stars lying between the two Aselli (ass-colts) which are imagined as feeding from it. It lies in the constellation Cancer.

We have now briefly recounted the observations made thus far with regard to the moon, the fixed stars, and the Milky Way. There remains the matter which in my opinion deserves to be considered the most important of all—the disclosure of four planets never seen from the creation of the world up to our own time, together with the occasion of my having discovered and studied them, their arrangements, and the observations made of their movements and alterations during the past two months. I invite all astronomers to apply themselves to examine them and determine their periodic times, something which has so far been quite impossible to complete, owing to the shortness of the time. Once more, however, warning is given that it will be necessary to have a very accurate telescope such as we have described at the beginning of this discourse.

On the seventh day of January in this present year 1610, at the first hour of night, when I was viewing the heavenly bodies with a telescope, Jupiter presented itself to me; and because I had prepared a very excellent instrument for myself, I perceived (as I had not before, on account of the weakness of my previous instrument) that beside the there were three starlets, small indeed, but very bright. Though I believed them to be among the host of fixed stars, they aroused my curiosity somewhat by appearing to lie in an exact straight line parallel to the ecliptic, and by their being more splendid than others of their size. Their arrangement with respect to Jupiter and each other was the following:

East   ✳   ✳   ◯   ✳   **West**

that is, there were two stars on the eastern side and one to the west. The most easterly star and the western one appeared larger than the other. I paid no attention to the distances between them and Jupiter, for at the outset I thought them to be fixed stars, as I have said.[20] But returning to the same investigation on January eighth—led by what, I do not know—I found a very different arrangement. The three starlets were now all to the west of Jupiter, closer together, and at equal intervals from one another as shown in the following sketch:

East   ○   ✳   ✳   ✳   **West**

At this time, though I did not yet turn my attention to the way the stars had come together, I began to concern myself with the question how Jupiter could be east of all these stars when on the previous day it had been west of two of them. I commenced to wonder whether Jupiter was not moving eastward at that time, contrary to the computations of the astronomers, and had got in front of them by that motion.[21] Hence it was with great interest that I awaited the next night. But I was disappointed in my hopes, for the sky was then covered with clouds everywhere.

On the tenth of January, however, the stars appeared in this position with respect to Jupiter:

East   ✳   ✳   ○   **West**

---

[20] The reader should remember that the telescope was nightly revealing to Galileo hundreds of fixed stars never previously observed. His unusual gifts for astronomical observation are illustrated by his having noticed and remembered these three merely by reason of their alignment, and recalling them so well that when by chance he happened to see them the following night he was certain that they had changed their positions. No such plausible and candid account of the discovery was given by the rival astronomer Simon Mayr, who four years later claimed priority. See pp. 233 ff. and note 4, 233-34.

[21] See notes 4, p. 12. Jupiter was at this time in "retrograde" motion; that is, the earth's motion made the planet appears to be moving westward among the fixed stars.

that is, there were but two of them, both easterly, the third (as I supposed) being hidden behind Jupiter. As at first, they were in the same straight line with Jupiter and were arranged precisely in the line of the zodiac. Noticing this, and knowing that there was no way in which such alterations could be attributed to Jupiter's motion, yet being certain that these were still the same stars I had observed (in fact no other was to be found along the line of the zodiac for a long way on either side of Jupiter), my perplexity was now transformed into amazement. I was sure that the apparent changes belonged not to Jupiter but to the observed stars, and I resolved to pursue this investigation with greater care and attention.

And thus, on the eleventh of January, I saw the following disposition:

East   ✳ ✳   ⬯   **West**

There were two stars, both to the east, the central one being three times as far from Jupiter as from the one farther east. The latter star was nearly double the size of the former, whereas on the night before they had appeared approximately equal.

I had now decided beyond all question that there existed in the heavens three stars wandering about Jupiter as do Venus and Mercury about the sun, and this became plainer than daylight from observations on similar occasions which followed. Nor were there just three such stars; four wanderers complete their revolutions about Jupiter, and of their alterations as observed more precisely later on we shall give a description here. Also I measured the distances between them by means of the telescope, using the method explained before. Moreover I recorded the times of the observations, especially when more than one was made during the same night—for the revolutions of these planets are so speedily completed that it is usually possible to take even their hourly variations.

Thus on the twelfth of January at the first hour of night I saw the stars arranged in this way:

East   ✳ ✳○ ✳   West

The most easterly star was larger than the western one, though both were easily visible and quite bright. Each was about two minutes of arc distant from Jupiter. The third star was invisible at first, but commenced to appear after two hours; it almost touched Jupiter on the east, and was quite small. All were on the same straight line directed along the ecliptic.

On the thirteenth of January four stars were seen by me for the first time, in this situation relative to Jupiter:

East   ✳ ○✳✳✳   West

Three were westerly and one was to the east; they formed a straight line except that the middle western star departed slightly toward the north. The eastern star was two minutes of arc away from Jupiter, and the intervals of the rest from one another and from Jupiter were about one minute. All the stars appeared to be of the same magnitude, and though small were very bright, much brighter than fixed stars of the same size.[22]

✳   ✳   ✳

On the twenty-sixth of February, midway in the first hour of night, there were only two stars:

East   ✳   ○   ✳   West

---

[22]   Galileo's day-by-day journal of observations continued in unbroken sequence until ten days before publication of the book, which he remained in Venice to supervise. The observations omitted here contained nothing of a novel character.

One was to the east, ten minutes from Jupiter; the other to the west, six minutes away. The eastern one was somewhat smaller than the western. But at the fifth hour three stars were seen:

East ✱ ⭕ ✳ ✴ *West*

In addition to the two already noticed, a third was discovered to the west near Jupiter; it had at first been hidden behind Jupiter and was now one minute away. The eastern one appeared farther away than before, being eleven minutes from Jupiter.

This night for the first time I wanted to observe the progress of Jupiter and its accompanying planets along the line of the zodiac in relation to some fixed star, and such a star was seen to the east, eleven minutes distant from the easterly starlet and a little removed toward the south, in the following manner:

East ✳ ⭕✴ ✴ *West*
★

On the twenty-seventh of February, four minutes after the first hour, the stars appeared in this configuration:

East ✴ ✴⭕ ✴✴ *West*
★

The most easterly was ten minutes from Jupiter; the next, thirty seconds; the next to the west was two minutes thirty seconds from Jupiter, and the most westerly was one minute from that. Those nearest Jupiter appeared very small, while the end ones were plainly visible, especially the westernmost. They marked out an exactly straight line along the course of the ecliptic. The progress of these planets toward the east is seen quite clearly by reference to the fixed star mentioned, since Jupiter and its accompanying planets were closer to it, as may be seen in the figure above. At the fifth hour, the eastern star closer to Jupiter was one minute away.

At the first hour on February twenty-eighth, two stars only were seen; one easterly, distant nine minutes from Jupiter, and one to the west, two minutes away. They were easily visible and on the same straight line. The fixed star, perpendicular to this line, now fell under the eastern as in this figure:

At the fifth hour a third star, two minutes east of Jupiter, was seen in this position:

On the first of March, forty minutes after sunset, four stars all to the east were seen, of which the nearest to Jupiter was two minutes away, the next was one minute from this, the third two seconds from that and brighter than any of the others; from this in turn the most easterly was four minutes distant, and it was smaller than the rest. They marked out almost a straight line, but the third one counting from Jupiter was a little to the north. The fixed star formed an equilateral triangle with Jupiter and the most easterly star, as in this figure:

On March second, half an hour after sunset, there were three planets, two to the east and one to the west, in this configuration:

The most easterly was seven minutes from Jupiter and thirty seconds from its neighbor; the western one was two minutes away from Jupiter. The end stars were very bright and were larger than that in the middle, which appeared very small. The most easterly star appeared a little elevated toward the north from the straight line through the other planets and Jupiter. The fixed star previously mentioned was eight minutes from the western planet along the line drawn from it perpendicularly to the straight line through all the planets, as shown above.

I have reported these relations of Jupiter and its companions with the fixed star so that anyone may comprehend that the progress of those planets, both in longitude and latitude, agrees exactly with the movements derived from planetary tables.

Such are the observations concerning the four Medicean planets recently first discovered by me, and although from these data their periods have not yet been reconstructed in numerical form, it is legitimate at least to put in evidence some facts worthy of note. Above all, since they sometimes follow and sometimes precede Jupiter by the same intervals, and they remain within very limited distances either to east or west of Jupiter, accompanying that planet in both its retrograde and direct movements in a constant manner, no one can doubt that they complete their revolutions about Jupiter and at the same time effect all together a twelve-year period about the center of the universe. That they also revolve in unequal circles is manifestly deduced from the fact that at the greatest elongation[23] from Jupiter it is never possible to see two of these planets in conjunction, whereas in the vicinity of Jupiter they are found united two, three, and sometimes all four together. It is also observed that the revolutions are swifter in those planets which describe smaller circles about Jupiter, since the stars closest to Jupiter are usually seen to the east when on the previous day they appeared to the west, and vice versa, while the planet which traces the largest orbit appears upon accurate observation of its returns to have a semimonthly period.

Here we have a fine and elegant argument for quieting the doubts of those who, while accepting with tranquil mind the revolutions of the planets about the sun in the Copernican system, are mightily disturbed to have the moon alone revolve about the earth and accompany it in an annual rotation about the sun. Some have believed that this structure of the universe should be rejected as impossible. But now we have not just one rotating about another while both run through a great orbit around the sun; our own eyes show us four stars which wander around Jupiter as does the moon around the earth, while all together trace out a grand revolution about the sun in the space of twelve years.

And finally we should not omit the reason for which the Medicean stars appear sometimes to be twice as large as at other times, though their orbits about Jupiter are very restricted. We certainly cannot seek the cause in terrestrial vapors, as Jupiter and its neighboring fixed stars are not seen to change size in the least while this increase and diminution are taking place. It is quite unthinkable that the cause of variation should be their change of distance from the earth at perigee and apogee, since a small circular rotation could by no means produce this effect, and an oval motion (which in

---

[23] By this is meant the greatest angular separation from Jupiter attained by any of the satellites.

this case would have to be nearly straight) seems unthinkable and quite inconsistent with the appearances.[24] But I shall gladly explain what occurs to me on this matter, offering it freely to the judgment and criticism of thoughtful men. It is known that the interposition of terrestrial vapors makes the sun and moon appear large, while the fixed stars and planets are made to appear smaller. Thus the two great luminaries are seen larger when close to the horizon, while the stars appear smaller and for the most part hardly visible. Hence the stars appear very feeble by day and in twilight, though the moon does not, as we have said. Now from what has been said above, and even more from what we shall say at greater length in our *system*, it follows that not only the earth but also the moon is surrounded by an envelope of vapors, and we may apply precisely the same judgment to the rest of the planets. Hence it does not appear entirely impossible to assume that around Jupiter also there exists an envelope denser than the rest of the aether, about which the Medicean planets revolve as does the moon about the elemental sphere. Through the interposition of this envelope they appear larger when they are in perigee by the removal, or at least the attenuation, of this envelope.

Time prevents my proceeding further, but the gentle reader may expect more soon.

FINIS

---

[24] The marked variation in brightness of the satellites which Galileo observed may be attributed mainly to markings upon their surfaces, though this was not determined until two centuries later. The mention here of a possible oval shape of the orbits is the closest Galileo ever came to accepting Kepler's great discovery of the previous year (cf. p. 17). Even here, however, he was probably not thinking of Kepler's work but of an idea proposed by earlier astronomers for the moon and the planet Venus.

GALILEO GALILEI

1564-1642

"Letter to the Grand Duchess Christina"

1616

Selection From:

Galilei, Galileo. 1616. "Letter to the Grand Duchess Christina." In *Discoveries and Opinions of Galileo*. Stillman Drake, Trans. New York: Doubleday Anchor Books. 1957. 21-58 and 175-202.

GALILEO GALILEI TO THE MOST SERENE GRAND DUCHESS MOTHER:

Some years ago, as Your Serene Highness well knows, I discovered in the heavens many things that had not been seen before our own age. The novelty of these things, as well as some consequences which followed from them in contradiction to the physical notions commonly held among academic philosophers, stirred up against me no small number of professors—as if I had placed these things in the sky with my own hands in order to upset nature and overturn the sciences. They seemed to forget that the increase of known truths stimulates the investigation, establishment, and growth of the arts; not their diminution or destruction.

Showing a greater fondness for their own opinions than for truth, they sought to deny and disprove the new things which, if they had cared to look for themselves, their own senses would have demonstrated to them. To this end they hurled various charges and published numerous writings filled with vain arguments, and they made the grave mistake of sprinkling these with passages taken from places in the Bible which they had failed to understand properly, and which were ill suited to their purposes.

These men would perhaps not have fallen into such error had they but paid attention to a most useful doctrine of St. Augustine's, relative to our making positive statements about things which are obscure and hard to understand by means of reason alone. Speaking of a certain physical conclusion about the heavenly bodies, he wrote: "Now keeping always our respect for moderation in grave piety, we ought not to believe anything inadvisedly on a dubious point, lest in favor to our error we conceive a prejudice against something that truth hereafter may reveal to be not contrary in any way to the sacred books of either the Old or the New Testament."[1]

Well, the passage of time has revealed to everyone the truths that I previously set forth; and, together with the truth of the facts, there has come to light the great difference in attitude between those who simply and dispassionately refused to admit the discoveries to be true, and those who combined with their incredulity some reckless passion of their own. Men who were well grounded in astronomical and physical science were persuaded as soon as they received my first message. There were others who denied them or remained in

---

[1] *De Genesi ad Literam,* end of bk ii (citations of theological works are taken from Galileo's notes, without verification.

doubt only because of their novel and unexpected character, and because they had not yet had the opportunity to see for themselves. These men have by degrees come to be satisfied. But some, besides allegiance to their original error, possess I know not what fanciful interest in remaining hostile not so much toward the things in question as toward their discoverer. No longer being able to deny them, these men now take refuge in obstinate silence, but being more than ever exasperated by that which has pacified and quieted other men, they divert their thoughts to other fancies and seek new ways to damage me.

I should pay no more attention to them than to those who previously contradicted me—at whom I always laugh, being assured of the eventual outcome—were it not that in their new calumnies and persecutions I perceive that they do not stop at proving themselves more learned than I am (a claim which I scarcely contest), but go so far as to cast against me imputations of crimes which must be, and are, more abhorrent to me than death itself. I cannot remain satisfied merely to know that the injustice of this is recognized by those who are acquainted with these men and with me, as perhaps it is not known to others.

Persisting in their original resolve to destroy me and everything mine by any mean they can think of, these men are aware of my views in astronomy and philosophy. They know that as to the arrangement of the parts of the universe, I hold the sun to be situated motionless in the center of the revolution of the celestial orbs while the earth rotates on its axis and revolves about the sun. They know also that I support this position not only by refuting the arguments of Ptolemy and Aristotle, but by producing many counterarguments; in particular, some which relate to physical effects whose causes can perhaps be assigned in no other way. In addition there are astronomical arguments derived from many things in my new celestial discoveries that plainly confute the Ptolemaic system while admirably agreeing with and confirming the contrary hypothesis. Possibly because they are disturbed by the known truth of other propositions of mine which differ from those commonly held, and therefore mistrusting their defense so long as they confine themselves to the field of philosophy, these men have resolved to fabricate a shield for their fallacies out of the mantle of pretended religion and the authority of the Bible. These they apply, with little judgment, to the refutation of arguments that they do not understand and have not even listened to.

First they have endeavored to spread the opinion that such propositions in general are contrary to the Bible and are consequently damnable and heretical. They know that it is human nature to take up causes whereby a man may oppress his neighbor, no matter how unjustly, rather than those from which a man may receive some just encouragement. Hence they have had no trouble in finding men who would preach the damnability and heresy of the new doctrine from their very pulpits with unwonted confidence, thus doing impious and inconsiderate injury not only to that doctrine and its followers but to all mathematics and mathematicians in general. Next, becoming bolder, and hoping (though vainly) that this seed which first took root in their hypocritical minds would send out branches and ascend to heaven, they began scattering rumors among the people that before long this doctrine would be condemned by the supreme authority. They know, too, that official condemnation would not only suppress the two propositions which I have mentioned, but would render damnable all other astronomical and physical statements and observations that have any necessary relation or connection with these.

In order to facilitate their designs, they seek so far as possible (at least among the common people) to make this opinion seem new and to belong to me alone. They pretend not to know that its author, or rather its restorer and confirmer, was Nicholas Copernicus; and that he was not only a Catholic, but a priest and a canon. He was in fact so esteemed by the church that when the Lateran Council under Leo X took up the correction of the church calendar, Copernicus was called to Rome from the most remote parts of Germany to undertake its reform. At that time the calendar was defective because the true measures of the year and the lunar month were not exactly known. The Bishop of Culm,[2] then superintendent of this matter, assigned Copernicus to seek more light and greater certainty concerning the celestial motions by means of constant study and labor. With Herculean toil he set his admirable mind to this task, and he made such great progress in this science and brought our knowledge of the heavenly motions to such precision that he became celebrated as an astronomer. Since that time not only has the calendar been regulated by his teachings, but tables of all the motions of the planets have been calculated as well.

Having reduced his system into six books, he published these at the instance of the Cardinal of Capua[3] and the Bishop of Culm. And since he had assumed his laborious enterprise by order of the supreme pontiff, he dedicated this book *On the celestial revolutions* to Pope Paul III. When printed, the book was accepted by the holy Church, and it has been read and studied by everyone without the faintest hint of any objection ever being conceived against its doctrines. Yet now that manifest experiences and necessary proofs have shown them to be well grounded, persons exist who would strip the author of his reward without so much as looking at his book, and add the shame of having him pronounced a heretic. All this they would do merely to satisfy their personal displeasure conceived without any cause against another man, who has no interest in Copernicus beyond approving his teachings.

---

[2] Tiedmann, Geise, to whom Copernicus referred in his preface as "that scholar, my good friend."

[3] Nicholas Schoenberg, spoken of by Copernicus as "celebrated in all fields of scholarship."

Now as to the false aspersions which they so unjustly seek to cast upon me, I have thought it necessary to justify myself in the eyes of all men, whose judgment in matters of religion and of reputation I must hold in great esteem. I shall therefore discourse of the particulars which these men produce to make this opinion detested and to have it condemned not merely as false but as heretical. To this end they make a shield of their hypocritical zeal for religion. They go about invoking the Bible, which they would have minister to their deceitful purposes. Contrary to the sense of the Bible and the intention of the holy Fathers, if I am not mistaken, they would extend such authorities until even in purely physical matters—where faith is not involved—they would have us altogether abandon reason and the evidence of our senses in favor of some biblical passage, though under the surface meaning of its words this passage may contain a different sense.

I hope to show that I proceed with much greater piety than they do, when I argue not against condemning this book, but against condemning it in the way they suggest—that is, without understanding it, weighing it, or so much as reading it. For Copernicus never discusses matters of religion or faith, nor does he use arguments that depend in any way upon the authority of sacred writings which he might have interpreted erroneously. He stands always upon physical conclusions pertaining to the celestial motions, and deals with them by astronomical and geometrical demonstrations, founded primarily upon sense experiences and very exact observations. He did not ignore the Bible, but he knew very well that if his doctrine were proved, then it could not contradict the Scriptures when they were rightly understood. And thus at the end of his letter of dedication, addressing the pope, he said:

"If there should chance to be any exegetes ignorant of mathematics who pretend to skill in that discipline, and dare to condemn and censure this hypothesis of mine upon the authority of some scriptural passage twisted to their purpose, I value them not, but disdain their unconsidered judgment.

For it is known that Lactantius—a poor mathematician though in other respects a worthy author—writes very childishly about the shape of the earth when he scoffs at those who affirm it to be a globe. Hence it should not seem strange to the ingenious if people of that sort should in turn deride me. But mathematics is written for mathematicians, by whom, if I am not deceived, these labors of mine will be recognized as contributing something to their domain, as also to that of the Church over which Your Holiness now reigns."[4]

Such are the people who labor to persuade us that an author like Copernicus may be condemned without being read, and who produce various authorities from the Bible, from theologians, and from Church Councils to make us believe that this is not only lawful but commendable. Since I hold these to be of supreme authority, I consider it rank temerity for anyone to contradict them—when employed according to the usage of the holy Church. Yet I do not believe it is wrong to speak out when there is reason to suspect that other men wish, for some personal motive, to produce and employ such authorities for purposes quite different from the sacred intention of the holy Church.

Therefore I declare (and my sincerity will make itself manifest) not only that I mean to submit myself freely and renounce any errors into which I may fall in this discourse through ignorance of matters pertaining to religion, but that I do not desire in these matters to engage in disputes with anyone, even on points that are disputable. My goal is this alone; that if, among errors that may abound in these considerations of a subject remote from my profession, there is anything that may be serviceable to the holy Church in making a decision concerning the Copernican system, it may be taken and utilized as seems best to the superiors. And if not, let my book be torn and burnt, as I neither intend nor pretend to gain from it any fruit that is not pious and Catholic. And though many of the

---

[4]   *De Revolutionibus* (Nuremberg, 1543), f.iiii.

things I shall reprove have been heard by my own ears, I shall freely grant to those who have spoken them that they never said them, if that is what they wish, and I shall confess myself to have been mistaken. Hence let whatever I reply be addressed not to them, but to whoever may have held such opinions.

The reason produced for condemning the opinion that the earth moves and the sun stands still is that in many places in the Bible one may read that the sun moves and the earth stands still. Since the Bible cannot err, it follows as a necessary consequence that anyone takes an erroneous and heretical position who maintains that the sun is inherently motionless and the earth movable.

With regard to this argument, I think in the first place that it is very pious to say and prudent to affirm that the holy Bible can never speak untruth—whenever its true meaning is understood. But I believe nobody will deny that it is often very abstruse, and may say things which are quite different from what its bare words signify. Hence in expounding the Bible if one were always to confine oneself to the unadorned grammatical meaning, one might fall into error. Not only contradictions and propositions far from true might thus be made to appear in the Bible, but even grave heresies and follies. Thus it would be necessary to assign to God feet, hands, and eyes, as well as corporeal and human affections, such as anger, repentance, hatred, and sometimes even the forgetting of things past and ignorance of those to come. These propositions uttered by the Holy Ghost were set down in that manner by the sacred scribes in order to accommodate them to the capacities of the common people, who are rude and unlearned. For the sake of those who deserve to be separated from the herd, it is necessary that wise expositors should produce the true senses of such passages, together with the special reasons for which they were set down in these words. This doctrine is so widespread and so definite with all theologians that it would be superfluous to adduce evidence for it.

Hence I think that I may reasonably conclude that whenever the Bible has occasion to speak of any physical conclusion (especially those which are very abstruse and hard to understand), the rule has been observed of avoiding confusion in the minds of the common people which would render them contumacious toward the higher mysteries. Now the Bible, merely to condescend to popular capacity, has not hesitated to obscure some very important pronouncements, attributing to God himself some qualities extremely remote from (and even contrary to) His essence. Who, then, would positively declare that this principle has been set aside, and the Bible has confined itself rigorously to the bare and restricted sense of its words, when speaking but casually of the earth, of water, of the sun, or of any other created thing? Especially in view of the fact that these things in no way concern the primary purpose of the sacred writings, which is the service of God and the salvation of souls—matters infinitely beyond the comprehension of the common people.

This being granted, I think that in discussions of physical problems we ought to begin not from the authority of scriptural passages, but from sense-experiences and necessary demonstrations; for the holy Bible and the phenomena of nature proceed alike from the divine Word, the former as the dictate of the Holy Ghost and the latter as the observant executrix of God's commands. It is necessary for the Bible, in order to be accommodated to the understanding of every man, to speak many things which appear to differ from the absolute truth so far as the bare meaning of the words is concerned. But Nature, on the other hand, is inexorable and immutable; she never transgresses the laws imposed upon her, or cares a whit whether her abstruse reasons and methods of operation are understandable to men. For that reason it appears that nothing physical which sense-experience sets before our eyes, or which necessary demonstrations prove to us, ought to be called in question (much less condemned) upon the testimony of biblical passages which may have some different meaning beneath their words. For the Bible is not chained in every expression to conditions as strict as those which govern all physical effects; nor is God any less excellently revealed in Nature's actions than in the sacred statements of the Bible. Perhaps this is what Tertullian meant by these words:

"We conclude that God is known first through Nature, and then again, more particularly, by doctrine; by Nature in His works, and by doctrine in His revealed word."[5]

From this I do not mean to infer that we need not have an extraordinary esteem for the passages of holy Scripture. On the contrary, having arrived at any certainties in physics, we ought to utilize these as the most appropriate aids in the true exposition of the Bible and in the investigation of those meanings which are necessarily contained therein, for these must be concordant with demonstrated truths. I should judge that the authority of the Bible was designed to persuade men of those articles and propositions which, surpassing all human reasoning, could not be made credible by science, or by any other means than through the very mouth of the Holy Spirit.

Yet even in those propositions which are not matters of faith, this authority ought to be preferred over that of all human writings which are supported only by bare assertions or probable arguments, and not set forth in a demonstrative way. This I hold to be necessary and proper to the same extent that divine wisdom surpasses all human judgment and conjecture.

But I do not feel obliged to believe that that same God who has endowed us with senses, reason, and intellect has intended to forgo their use and by some other means to give us knowledge which we can attain by them. He would not require us to deny sense and reason in physical matters which are set before our eyes and minds by direct experience or necessary demonstrations. This must be especially true in those sciences of which but the faintest trace (and that consisting of conclusions) is to be found in the Bible. Of astronomy, for instance, so little is found that none of the planets except Venus are so much as mentioned, and this only once or twice under the name of

---

[5]   *Adversus Marcionem*, ii, 18.

"Lucifer." If the sacred scribes had had any intention of teaching people certain arrangements and motions of the heavenly bodies, or had they wished us to derive such knowledge from the Bible, then in my opinion they would not have spoken of these matters so sparingly in comparison with the infinite number of admirable conclusions which are demonstrated in that science. Far from pretending to teach us the constitution and motions of the heavens and the stars, with their shapes, magnitudes, and distances, the authors of the Bible intentionally forbore to speak of these things, though all were quite well known to them. Such is the opinion of the holiest and most learned Fathers, and in St. Augustine we find the following words:

"It is likewise commonly asked what we may believe about the form and shape of the heavens according to the Scriptures, for many contend much about these matters. But with superior prudence our authors have forborne to speak of this, as in no way furthering the student with respect to a blessed life—and, more important still, as taking up much of that time which should be spent in holy exercises. What is it to me whether heaven, like a sphere, surrounds the earth on all sides as a mass balanced in the center of the universe, or whether like a dish it merely covers and overcasts the earth? Belief in Scripture is urged rather for the reason we have often mentioned; that is, in order that no one, through ignorance of divine passages, finding anything in our Bibles or hearing anything cited from them of such a nature as may seem to oppose manifest conclusions, should be induced to suspect their truth when they teach, relate, and deliver more profitable matters. Hence let it be said briefly, touching the form of heaven, that our authors knew the truth but the Holy Spirit did not desire that men should learn things that are useful to no one for salvation."[6]

The same disregard of these sacred authors toward beliefs about the phenomena of the celestial bodies is repeated to us

---

[6]   De Genesi ad Literam ii, 9. Galileo has noted also: "The same is to be read in Peter the Lombard, master of opinions."

by St. Augustine in his next chapter. On the question whether we are to believe that the heaven moves or stands still, he writes thus:

"Some of the brethren raise a question concerning the motion of heaven, whether it is fixed or moved. If it is moved, they say, how is it a firmament? If it stands still, how do these stars which are held fixed in it go round from east to west, the more northerly performing shorter circuits near the pole, so that heaven (if there is another pole unknown to us) may seem to revolve upon some axis, or (if there is no other pole) may be thought to move as a discus? To these men I reply that it would require many subtle and profound reasonings to find out which of these things is actually so; but to undertake this and discuss it is consistent neither with my leisure nor with the duty of those whom I desire to instruct in essential matters more directly conducing to their salvation and to the benefit of the holy Church."[7]

From these things it follows as a necessary consequence that, since the Holy Ghost did not intend to teach us whether heaven moves or stands still, whether its shape is spherical or like a discus or extended in a plane, nor whether the earth is located at its center or off to one side, then so much the less was it intended to settle for us any other conclusion of the same kind. And the motion or rest of the earth and the sun is so closely linked with the things just named, that without a determination of the one, neither side can be taken in the other matters. Now if the Holy Spirit has purposely neglected to teach us propositions of this sort as irrelevant to the highest goal (that is, to our salvation), how can anyone affirm that it is obligatory to take sides on them, and that one belief is required by faith, while the other side is erroneous? Can an opinion be heretical and yet have no concern with the salvation of souls? Can the Holy Ghost be asserted not to have intended teaching us something that does concern our salvation? I would say here something that was heard from an ecclesiastic of the most

---

7   *Ibid.*, ii, 10.

eminent degree: "That the intention of the Holy Ghost is to teach us how one goes to heaven, not how heaven goes."[8]

But let us again consider the degree to which necessary demonstrations and sense experiences ought to be respected in physical conclusions, and the authority they have enjoyed at the hands of holy and learned theologians. From among a hundred attestations I have selected the following:

"We must also take heed, in handling the doctrine of Moses, that we altogether avoid saying positively and confidently anything which contradicts manifest experiences and the reasoning of philosophy or the other sciences. For since every truth is in agreement with all other truth, the truth of Holy Writ cannot be contrary to the solid reasons and experiences of human knowledge."[9]

And in St. Augustine we read: "If anyone shall set the authority of Holy Writ against clear and manifest reason, he who does this knows not what he has undertaken; for he opposes to the truth not the meaning of the Bible, which is beyond his comprehension, but rather his own interpretation; not what is in the Bible, but what he has found in himself and imagines to be there."[10]

This granted, and it being true that two truths cannot contradict one another, it is the function of wise expositors to seek out the true senses of scriptural texts. These will unquestionably accord with the physical conclusions which manifest sense and necessary demonstrations have previously made certain to us. Now the Bible, as has been remarked, admits in many places expositions that are remote from the signification of the words for reasons we have already given. Moreover, we are unable to affirm that all interpreters of the Bible speak by divine inspiration, for if that were so there would exist no differences between them about the sense of a given

---

[8] A marginal note by Galileo assigns this epigram to Cardinal Baronius (1538-1607). Baronius visited Padua with Cardinal Bellarmine in 1598, and Galileo probably met him at that time.

[9] Pererius on Genesis, near the beginning.

[10] In the seventh letter to Marcellinus.

passage. Hence I should think it would be the part of prudence not to permit anyone to usurp scriptural texts and force them in some way to maintain any physical conclusion to be true, when at some future time the senses and demonstrative or necessary reasons may show the contrary. Who indeed will set bounds to human ingenuity? Who will assert that everything in the universe capable of being perceived is already discovered and known? Let us rather confess quite truly that "Those truths which we know are very few in comparison with those which we do not know."

We have it from the very mouth of the Holy Ghost that God delivered up the world to disputations, *so that man cannot find out the work that God hath done from the beginning even to the end.*[11] In my opinion no one, in contradiction to that dictum, should close the road to free philosophizing about mundane and physical things, as if everything had already been discovered and revealed with certainty. Nor should it be considered rash not to be satisfied with those opinions which have become common. No one should be scorned in physical disputes for not holding to the opinions which happen to please other people best, especially concerning problems which have been debated among the greatest philosophers for thousands of years. One of these is the stability of the sun and mobility of the earth, a doctrine believed by Pythagoras and all his followers, by Heracleides of Pontus[12] (who was one of them), by Philolaus the teacher of Plato,[13] and

---

[11] Ecclesiastes 3:11.

[12] Heracleides was born about 390 B.C. and is said to have attended lectures by Aristotle of Athens. He believed that the earth rotated on its axis, but not that it moved around the sun. He also discovered that Mercury and Venus revolve around the sun, and may have developed a system similar to that of Tycho.

[13] Philolaus, an early follower of Pythagoras, flourished at Thebes toward the end of the fifth century B.C. Although a contemporary of Socrates, the teacher of Plato, he had nothing to do with Plato's instruction. According to Philolaus the earth revolved around a central fire, but not about the sun (cf. note 7, p. 34).

by Plato himself according to Aristotle. Plutarch writes in his *Life of Numa* that Plato, when he had grown old, said it was most absurd to believe otherwise.[14] The same doctrine was held by Aristarchus of Samos,[15] as Archimedes tells us; by Seleucus[16] the mathematician, by Nicetas[17] the philosopher (on the testimony of Cicero), and by many others. Finally this opinion has been amplified and confirmed with many observations and demonstrations by Nicholas Copernicus. And Seneca,[18] a most eminent philosopher, advises us in his book on comets that we should more diligently seek to ascertain whether it is in the sky or in the earth that the diurnal rotation resides.

Hence it would probably be wise and useful counsel if, beyond articles which concern salvation and the establishment

---

[14] "Plato held opinion in that age, that the earth was in another place than in the very middest, and that the centre of the world, as the most honourable place, did appertain to some other of more worthy substance than the earth." (Trans. Sir Thomas North.) This translation is no longer accepted.

[15] Aristarchus (ca. 310-230 B.C.) was the true forerunner of Copernicus in antiquity, and not the Pythagoreans as was generally believed in Galileo's time.

[16] Seleucus, who flourished about 150 B.C., is the only ancient astronomer known to have adopted the heliocentric system of Aristarchus. After his time this gave way entirely to the system founded by his contemporary Hipparchus.

[17] Nicetus is an incorrect form given by Copernicus to the name of Hicetas of Syracuse. Of this mathematician nothing is known beyond the fact that some of the ancients credited him instead of Philolaus with the astronomy which came to be associated with the Pythagoreans in general.

[18] Seneca (ca. 3-65 A.D.) was the tutor to Nero. He devoted the seventh book of his *Quaestiones Naturales* to comets. In the second chapter of this book he raised the question of the earth's rotation, and in the final chapters he appealed for patience and further investigation into such matters.

of our Faith, against the stability of which there is no danger whatever that any valid and effective doctrine can ever arise, men would not aggregate further articles unnecessarily. And it would certainly be preposterous to introduce them at the request of persons who, besides not being known to speak by inspiration of divine grace, are clearly seen to lack that understanding which is necessary in order to comprehend, let alone discuss, the demonstrations by which such conclusions are supported in the subtler sciences. If I may speak my opinion freely, I should say further that it would perhaps fit in better with the decorum and majesty of the sacred writings to take measures for preventing every shallow and vulgar writer from giving to his compositions (often grounded upon foolish fancies) an air of authority by inserting in them passages from the Bible, interpreted (or rather distorted) into senses as far from the right meaning of Scripture as those authors are near to absurdity who thus ostentatiously adorn their writings. Of such abuses many examples might be produced, but for the present I shall confine myself to two which are germane to these astronomical matters. The first concerns those writings which were published against the existence of the Medicean planets recently discovered by me, in which many passages of holy Scripture were cited.[19] Now that everyone has seen these planets, I should like to know what new interpretations those

---

[19]   The principle book which had offended in this regard was the *Dianoia Astronomica . . .* of Francesco Sizzi (Venice, 1611). About the time Galileo arrived at Florence, Sizzi departed for France, where he came into association with some good mathematicians. In 1613 he wrote to a friend at Rome to express his admiration of Galileo's work on floating bodies and to deride its opponents. The letter was forwarded to Galileo. In it Sizzi had reported, though rather cryptically, upon some French observations concerning sunspots, and it was probably this which led Galileo to his knowledge of the tilt sun's axis (cf. note 14, p. 125). Sizzi was broken on the wheel in 1617 for writing a pamphlet against the king of France.

same antagonists employ in expounding the Scripture and excusing their own simplicity. My other example is that of a man who has lately published, in defiance of astronomers and philosophers, the opinion that the moon does not receive its light from the sun but is brilliant by its own nature.[20] He supports this fancy (or rather thinks he does) by sundry texts of Scripture which he believes cannot be explained unless his theory is true; yet that the moon is inherently dark is surely as plain as daylight.

It is obvious that such authors, not having penetrated the true senses of Scripture, would impose upon others an obligation to subscribe to conclusions that are repugnant to manifest reason and sense, if they had any authority to do so. God forbid that this sort of abuse should gain countenance and authority, for then in a short time it would be necessary to proscribe all the contemplative sciences. People who are unable to understand perfectly both the Bible and the sciences far outnumber those who do understand. The former, glancing superficially through the Bible, would arrogate to themselves the authority to decree upon every question of physics on the strength of some word which they have misunderstood, and which was employed by the sacred authors for some different purpose. And the smaller number of understanding men could not dam up the furious torrent of such people, who would gain the majority of followers simply because it is much more pleasant to gain a reputation for wisdom without effort or study than to consume oneself tirelessly in the most laborious disciplines. Let us therefore render thanks to Almighty God, who in His beneficence protects us from this

---

[20] This is frequently said to refer to J.C. Lagalla's *De phaenominis in orbe lunae* . . . (Venice, 1612), a wretched book which has the sole distinction of being the first to mention the word "telescope" in print. A more probable reference, however, seems to be to the *Dialogo di Fr. Ulisse Albergotti* . . . *nel quale si tiene* . . . *la Luna esser da sé luminosa* . . . (Viterbo, 1613).

danger by depriving such persons of all authority, reposing the power of consultation, decision, and decree on such important matters in the high wisdom and benevolence of most prudent Fathers, and in the supreme authority of those who cannot fail to order matters properly under the guidance of the Holy Ghost. Hence we need not concern ourselves with the shallowness of those men whom grave and holy authors rightly reproach, and of whom in particular St. Jerome said, in reference to the Bible:

"This is ventured upon, lacerated, and taught by the garrulous old woman, the doting old man, and the prattling sophist before they have learned it. Others, led on by pride, weigh heavy words and philosophize amongst women concerning holy Scripture. Others—oh, shame!—learn from women what they teach to men, and (as if that were not enough) glibly expound to others that which they themselves do not understand. I forbear to speak of those of my own profession who, attaining a knowledge of the holy Scriptures after mundane learning, tickle the ears of the people with affected and studied expressions, and declare that everything they say is to be taken as the law of God. Not bothering to learn what the prophets and the apostles have maintained, they wrest incongruous testimonies into their own senses—as if distorting passages and twisting the Bible to their individual and contradictory whims were the genuine way of teaching, and not a corrupt one."[21]

I do not wish to place in the number of such lay writers some theologians whom I consider men of profound learning and devout behavior, and who are therefore held by me in great esteem and veneration. Yet I cannot deny that I feel some discomfort which I should like to have removed, when I hear them pretend to the power of constraining others by scriptural authority to follow in a physical dispute that opinion

---

[21] *Epistola ad Paulinum,* 103.

which they think best agrees with the Bible, and then believe themselves not bound to answer the opposing reasons and experiences. In explanation and support of this opinion they say that since theology is queen of all the sciences, she need not bend in any way to accommodate herself to the teachings of less worthy sciences which are subordinate to her; these others must rather be referred to her as to their supreme empress, changing and altering their conclusions according to her statutes and decrees. They add further that if in the inferior sciences any conclusion should be taken as certain in virtue of demonstrations or experiences, while in the Bible another conclusion is found repugnant to this, then the professors of that science should themselves undertake to undo their proofs and discover the fallacies in their own experiences, without bothering the theologians and exegetes. For, they say, it does not become the dignity of theology to stoop to the investigation of fallacies in the subordinate sciences; it is sufficient for her merely to determine the truth of a given conclusion with absolute authority, secure in her inability to err.

Now the physical conclusions in which they say we ought to be satisfied by Scripture, without glossing or expounding it in senses different from the literal, are those concerning which the Bible always speaks in the same manner and which the holy Fathers all receive and expound in the same way. But with regard to these judgments I have had occasion to consider several things, and I shall set them forth in order that I may be corrected by those who understand more than I do in these matters—for to their decisions I submit at all times.

First, I question whether there is not some equivocation in failing to specify the virtues which entitle sacred theology to the title of "queen." It might deserve that name by reason of including everything that is learned from all the other sciences and establishing everything by better methods and with profounder learning. It is thus, for example, that the rules for measuring fields and keeping accounts are much more excellently contained in arithmetic and in the geometry of Euclid than in the practices of surveyors and accountants. Or theology might be queen because of being occupied with a subject which excels in dignity all the subjects which compose the other sciences, and because her teachings are divulged in more sublime ways.

That the title and authority of queen belongs to theology in the first sense, I think will not be affirmed by theologians who have any skill in the other sciences. None of these, I think, will say that geometry, astronomy, music, and medicine are much more excellently contained in the Bible than they are in the books of Archimedes, Ptolemy, Boethius, and Galen. Hence it seems likely that regal pre-eminence is given to theology in the second sense; that is, by reason of its subject and the miraculous communication of divine revelation of conclusions which could not be conceived by men in any other way, concerning chiefly the attainment of eternal blessedness.

Let us grant then that theology is conversant with the loftiest divine contemplation, and occupies the regal throne among sciences by dignity. But acquiring the highest authority in this way, if she does not descend to the lower and humbler speculations of the subordinate sciences and has no regard for them because they are not concerned with blessedness, then her professors should not arrogate to themselves the authority to decide on controversies in professions which they have neither studied nor practiced. Why, this would be as if an absolute despot, being neither a physician nor an architect but knowing himself free to command, should undertake to administer medicines and erect buildings according to his whim—at grave peril of his poor patients' lives, and the speedy collapse of his edifices.

Again, to command that the very professors of astronomy themselves see to the refutation of their own observations and proofs as mere fallacies and sophisms is to enjoin something that lies beyond any possibility of accomplishment. For this would amount to commanding that they must not see what they see and must not understand what they know, and that in searching they must find the opposite of what they actually encounter. Before this could be done they would have to be taught how to make one mental faculty command another, and the inferior powers the superior, so that the imagination and the will might be forced to believe the opposite of what the intellect understands. I am referring at all times to merely physical propositions, and not to supernatural things which are matters of faith.

I entreat those wise and prudent Fathers to consider with great care the difference that exists between doctrines subject to proof and those subject to opinion. Considering the force exerted by logical deductions, they may ascertain that it is not in the power of the professors of demonstrative sciences to change their opinions at will and apply themselves first to one side and then to the other. There is a great difference between commanding a mathematician or a philosopher and influencing a lawyer or a merchant, for demonstrated conclusions about things in nature or in the heavens cannot be changed with the same facility as opinions about what is or is not lawful in a contract, bargain, or bill of exchange. This difference was well understood by the learned and holy Fathers, as proven by their having taken great pains in refuting philosophical fallacies. This may be found expressly in some of them; in particular, we find the following words of St. Augustine: "It is to be held as an unquestionable truth that whatever the sages of this world have demonstrated concerning physical matters is in no way contrary to our Bibles; hence whatever the sages teach in their books that is contrary to the holy Scriptures may be concluded without any hesitation to be quite false. And according to our ability let us make this evident, and let us keep the faith of our Lord, in whom are hidden all the treasures of wisdom, so that we neither become seduced by the verbiage of false philosophy nor frightened by the superstition of counterfeit religion."[22]

From the above words I conceive that I may deduce this doctrine: That in the books of the sages of this world

---

[22] *De Genesi ad literam* i, 21.

there are contained some physical truths which are soundly demonstrated, and others that are merely stated; as to the former, it is the office of wise divines to show that they do not contradict the holy Scriptures. And as to the propositions which are stated but not rigorously demonstrated, anything contrary to the Bible involved by them must be held undoubtedly false and should be proved so by every possible means.

Now if truly demonstrated physical conclusions need not be subordinated to biblical passages, but the latter must rather be shown not to interfere with the former, then before a physical proposition is condemned it must be shown to be not rigorously demonstrated—and this is to be done not by those who hold the proposition to be true, but by those who judge it to be false. This seems very reasonable and natural, for those who believe an argument to be false may much more easily find the fallacies in it than men who consider it to be true and conclusive. Indeed, in the latter case it will happen that the more the adherents of an opinion turn over their pages, examine the arguments, repeat the observations, and compare the experiences, the more they will be confirmed in that belief. And Your Highness knows what happened to the late mathematician of the University of Pisa[23] who undertook in his old age to look into the Copernican doctrine in the hope of shaking its foundations and refuting it, since he considered it false only because he had never studied it. As it fell out, no sooner had he understood its grounds, procedures, and demonstrations than he found himself persuaded, and from an opponent he became a very staunch defender of it. I might also name other mathematicians[24] who, moved by my latest discoveries, have confessed it necessary to alter the previously accepted system of the world, as this is simply unable to subsist any longer.

If in order to banish the opinion in question from the world it were sufficient to stop the mouth of a single man—as

---

[23]   Antonia Santucci (d. 1613).

[24]   A marginal note by Galileo here mentions Father Clavius; cf. p. 153.

perhaps those men persuade themselves who, measuring the minds of others by their own, think it impossible that this doctrine should be able to continue to find adherents—then that would be very easily done. But things stand otherwise. To carry out such a decision it would be necessary not only to prohibit the book of Copernicus and the writings of other authors who follow the same opinion, but to ban the whole science of astronomy. Furthermore, it would be necessary to forbid men to look at the heavens, in order that they might not see Mars and Venus sometimes quite near the earth and sometimes very distant, the variation being so great that Venus is forty times and Mars sixty times as large at one time as another. And it would be necessary to prevent Venus being seen round at one time and forked at another, with very thin horns; as well as many other sensory observations which can never be reconciled with the Ptolemaic system in any way, but are very strong arguments for the Copernican. And to ban Copernicus now that his doctrine is daily reinforced by many new observations and by the learned applying themselves to the reading of his book, after this opinion has been allowed and tolerated for those many years during which it was less followed and less confirmed, would seem in my judgment to be a contravention of truth, and an attempt to hide and suppress her the more as she revealed herself the more clearly and plainly. Not to abolish and censure his whole book, but only to condemn as erroneous this particular proposition, would (if I am not mistaken) be a still greater detriment to the minds of men, since it would afford them occasion to see a proposition proved that it was heresy to believe. And to prohibit the whole science would be but to censure a hundred passages of holy Scripture which teach us that the glory and greatness of Almighty God are marvelously discerned in all his works and divinely read in the open book of heaven. For let no one believe that reading the lofty concepts written in that book leads to nothing further than the mere seeing of the splendor of the sun and the stars and their rising and setting, which is as far as the eyes of brutes and of the vulgar can penetrate. Within its pages are couched mysteries so profound and concepts so sublime that the vigils, labors, and studies of hundreds upon hundreds of the most acute minds have still not pierced them, even after continual investigations for thousands of years. The eyes of an idiot perceive little by beholding the external appearance of a human body, as compared with the wonderful contrivances which a careful and practiced anatomist or philosopher discovers in that same body when he seeks out the use of all those muscles, tendons, nerves, and bones; or when examining the functions of the heart and the other principal organs, he seeks the seat of the vital faculties, notes and observes the admirable structure of the sense organs, and (without ever ceasing in his amazement and delight) contemplates the receptacles of the imagination, the memory, and the understanding. Likewise, that which presents itself to mere sight is as nothing in comparison with the high marvels that the ingenuity of learned men discovers in the heavens by long and accurate observation. And that concludes what I have to say on this matter.

Next let us answer those who assert that those physical propositions of which the Bible speaks always in one way, and which the Fathers all harmoniously accept in the same sense, must be taken according to the literal sense of the words without glosses or interpretations, and held as most certain and true. The motion of the sun and stability of the earth, they say, is of this sort; hence it is a matter of faith to believe in them, and the contrary view is erroneous.

To this I wish first to remark that among physical propositions there are some with regard to which all human science and reason cannot supply more than a plausible opinion and a probable conjecture in place of a sure and demonstrated knowledge; for example, whether the stars are animate. Then there are other propositions of which we have (or may confidently expect) positive assurances through experiments, long observation, and rigorous demonstration;

for example, whether or not the earth and the heavens move, and whether or not the heavens are spherical. As to the first sort of propositions, I have no doubt that where human reasoning cannot reach—and where consequently we can have no science but only opinion and faith—it is necessary in piety to comply absolutely with the strict sense of Scripture. But as to the other kind, I should think, as said before, that first we are to make certain of the fact, which will reveal to us the true senses of the Bible, and these will most certainly be found to agree with the proved fact (even though at first the words sounded otherwise), for two truths can never contradict each other. I take this to be an orthodox and indisputable doctrine, and I find it specifically in St. Augustine when he speaks of the shape of heaven and what we may believe concerning that. Astronomers seem to declare what is contrary to Scripture, for they hold the heavens to be spherical, while the Scripture calls it "stretched out like a curtain."[25] St. Augustine opines that we are not to be concerned lest the Bible contradict astronomers; we are to believe its authority if what they say is false and is founded only on the conjectures of frail humanity. But if what they say is proved by unquestionable arguments, this holy Father does not say that the astronomers are to be ordered to dissolve their proofs and declare their own conclusions to be false. Rather, he says it must be demonstrated that what is meant in the Bible by "curtain" is not contrary to their proofs. Here are his words:

"But some raise the following objection. 'How is it that the passage in our Bibles, *Who stretcheth out the heavens as a curtain,* does not contradict those who maintain the heavens to have a spherical shape?' It does contradict them if what they affirm is false, for that is true which is spoken by divine authority rather than that which proceeds from human frailty. But if, peradventure, they should be able to prove their position by experiences which place it beyond question, then it is to

---

[25]   Psalms 103:2 (Douay); 104:2 (King James).

be demonstrated that our speaking of a curtain in no way contradicts their manifest reasons."[26]

He then proceeds to admonish us that we must be no less careful and observant in reconciling a passage of the Bible with any demonstrated physical proposition than with some other biblical passage which might appear contrary to the first. The circumspection of this saint indeed deserves admiration and imitation, when even in obscure conclusions (of which we surely can have no knowledge through human proofs) he shows great reserve in determining what is to be believed. We see this from what he writes at the end of the second book of his commentary on Genesis, concerning the question whether the stars are to be believed animate:

"Although at present this matter cannot be settled, yet I suppose that in our further dealing with the Bible we may meet with other relevant passages, and then we may be permitted, if not to determine anything finally, at least to gain some hint concerning this matter according to the dictates of sacred authority. Now keeping always our respect for moderation in grave piety, we ought not to believe anything inadvisedly on a dubious point, lest in favor of our error we conceive a prejudice against something that truth hereafter may reveal to be not contrary in any way to the sacred books of either the Old or the New Testament."

From this and other passages the intention of the holy Fathers appears to be (if I am not mistaken) that in questions of nature which are not matters of faith it is first to be considered whether anything is demonstrated beyond doubt or known by sense-experience, or whether such knowledge or proof is possible; if it is, then, being the gift of God, it ought to be applied to find out the true senses of holy Scripture in those passages which superficially might seem to declare differently. These senses would unquestionably be discovered by wise theologians,

---

[26]   *De Genesi ad literam* [ii,] 9.

together with the reasons for which the Holy Ghost sometimes wished to veil itself under words of different meaning, whether for our exercise, or for some purpose unknown to me.

As to the other point, if we consider the primary aim of the Bible, I do not think that its having always spoken in the same sense need disturb this rule. If the Bible, accommodating itself to the capacity of the common people, has on one occasion expressed a proposition in words of different sense from the essence of that proposition, then why might it not have done the same, and for the same reason, whenever the same thing happened to be spoken of? Nay, to me it seems that not to have done this would but have increased confusion and diminished belief among the people.

Regarding the state of rest or motion of the sun and earth, experience plainly proves that in order to accommodate the common people it was necessary to assert of these things precisely what the words of the Bible convey. Even in our own age, people far less primitive continue to maintain the same opinion for reasons which will be found extremely trivial if well weighed and examined, and upon the basis of experiences that are wholly false or altogether beside the point. Nor is it worth while to try to change their opinion, they being unable to understand the arguments on the opposite side, for these depend upon observations too precise and demonstrations too subtle, grounded on abstractions which require too strong an imagination to be comprehended by them. Hence even if the stability of heaven and the motion of the earth should be more than certain in the minds of the wise, it would still be necessary to assert the contrary for the preservation of belief among the all-too-numerous vulgar. Among a thousand ordinary men who might be questioned concerning these things, probably not a single one will be found to answer anything except that it looks to him as if the sun moves and the earth stands still, and therefore he believes this to be certain. But one need not on that account take the common popular assent as an argument for the truth of what is stated; for if we should examine these very men concerning their reasons for what they believe, and on the other hand listen to the experiences and proofs which induce a few others to believe the contrary, we should find the latter to be persuaded by very sound arguments, and the former by simple appearances and vain or ridiculous impressions.

It is sufficiently obvious that to attribute motion to the sun and rest to the earth was therefore necessary lest the shallow minds of the common people should become confused, obstinate, and contumacious in yielding assent to the principal articles that are absolutely matters of faith. And if this was necessary, there is no wonder at all that it was carried out with great prudence in the holy Bible. I shall say further that not only respect for the incapacity of the vulgar, but also current opinion in those times, made the sacred authors accommodate themselves (in matters unnecessary to salvation) more to accepted usage than to the true essence of things. Speaking of this, St. Jerome writes:

"As if many things were not spoken in the Holy Bible according to the judgment of those times in which they were acted, rather than according to the truth contained."[27] And elsewhere the same saint says: "It is the custom for the biblical scribes to deliver their judgments in many things according to the commonly received opinion of their times."[28] And on the words in the twenty-sixth chapter of Job, *He stretcheth out the north over the void, and hangeth the earth above* nothing,[29] St. Thomas Aquinas notes that the Bible calls "void" or "nothing" that space which we know to be not empty, but filled with air. Nevertheless the Bible, he says, in order to accommodate itself to the beliefs of the common people (who think there is nothing in that space), calls it "void" and "nothing." Here are the words of St. Thomas: "What appears to us in the upper hemisphere of the heavens to be empty, and not a space filled with air, the common people regard as void; and it is usually

---

[27]   On Jeremiah, ch. 28.

[28]   On Matthew, ch. 13.

[29]   Job 26:7

spoken of in the holy Bible according to the ideas of the common people."[30]

Now from this passage I think one may very logically argue that for the same reason the Bible had still more cause to call the sun movable and the earth immovable. For if we were to test the capacity of the common people, we should find them even less apt to be persuaded of the stability of the sun and the motion of the earth than to believe that the space which environs the earth is filled with air. And if on this point it would not have been difficult to convince the common people, and yet the holy scribes forbore to attempt it, then it certainly must appear reasonable that in other and more abstruse propositions they have followed the same policy.

Copernicus himself knew the power over our ideas that is exerted by custom and by our inveterate way of conceiving things since infancy. Hence, in order not to increase for us the confusion and difficulty of abstraction, after he had first demonstrated that the motions which appear to us to belong to the sun or to the firmament are really not there but in the earth, he went on calling them motions of the sun and of the heavens when he later constructed his tables to apply them to use. He thus speaks of "sunrise" and "sunset," of the "rising and setting" of the stars, of changes in the obliquity of the ecliptic and of variations in the equinoctial points, of the mean motion and variations in motion of the sun, and so on. All these things really relate to the earth, but since we are fixed to the earth and consequently share in its every motion, we cannot discover them in the earth directly, and are obliged to refer them to the heavenly bodies in which they make their appearance to us. Hence we name them as if they took place where they appear to us to take place; and from this one may see how natural it is to accommodate things to our customary way of seeing them . . . .

---

[30] Aquinas on Job.

ISAAC NEWTON

1643-1727

Principia Mathematica

1729

Sir Isaac Newton is considered by many to be one of the greatest scientists and mathematicians of all time. Noted for his pioneering work in creating three basic laws of motion and gravity, he laid the groundwork for mechanics as well as for differential and integral calculus. His interests included studies in mathematics, optics, and astronomy, as well as alchemy, the forerunner of modern day chemistry.

Although not considered a promising student as a child, he enrolled in Trinity College Cambridge where he received his degree in 1665. He was appointed to a teaching position at Cambridge but the school was soon closed for nearly two years because of the plague. It was during those two years at home that he developed his thinking in mathematics, mechanics, and optics. His first text, *De Methodis Serierum et Fluxionun* was written in 1671, but remained unpublished until 1736. He later received a master's degree from Cambridge followed by a major fellowship in 1668. In 1670, he succeeded his professor and mentor, Isaac Barrow, as the second Lucasian Professor of Mathematics at Cambridge. He left Cambridge in 1696 to become Warden of the Royal Mint, a position that could have been mostly ceremonial. However, he took those duties seriously and designed new coinage, helped the government to move from silver to a gold standard, and developed ways to prevent counterfeiting. In 1703, he was elected president of the Royal Society and was so elected annually for the remainder of his life. He was knighted by Queen Anne for his many scientific contributions and was the first scientist to be so recognized.

Newton and Gottfried Wilhelm von Leibniz are both credited with developing calculus, as it appears that they worked independently. Although Newton invented calculus before Leibniz, he did not publish his work until after Leibniz published his. Who invented calculus led to a long and bitter debate between English and continental mathematicians, which lasted for many years, and invokes differing opinions to this day.

A deeply religious man, Newton first planned to study for theology and remained a true believer in the necessity of a God whom he thought designed and maintains the universe on its natural track. "His theological views are characterized by his belief that the beauty and regularity of the natural world could only 'proceed from the counsel and dominion of an intelligent and powerful Being.' He felt that 'the Supreme God exists necessarily, and by the same necessity he exists always and everywhere.'"[1]

Sources:

"Sir Isaac Newton." 2000. School of Mathematics and Statistics, University of St. Andrews, Scotland. Retrieved February 12, 2007, from http://www.gap-system.org/~history/Biographies/Newton.html.

Weisstein, Eric. 2007. "Newton, Isaac." Wolfram Research. Retrieved Feb 12, 2007, from http://scienceworld.wolfram.com/biography/Newton.html.

---

[1]   http://scienceworld.wolfram.com/biography/Newton.html

Selection From:

Newton, Isaac. 1687. *"Philosophiae naturalis principia mathematica." Mathematical Principles of Natural Philosophy and His System of the World.* Trans. Andrew Motte, 1729. Berkeley: University of California Press. 1934. xvii-xviii, 6-14, 398-400, 543-545.

PREFACE TO THE FIRST EDITION

SINCE THE ANCIENTS (as we are told by *Pappus*) esteemed the science of mechanics of greatest importance in the investigation of natural things, and the moderns, rejecting substantial forms and occult qualities, have endeavored to subject the phenomena of nature to the laws of mathematics, I have in this treatise cultivated mathematics as far as it relates to philosophy. The ancients considered mechanics in a twofold respect; as rational, which proceeds accurately by demonstration, and practical. To practical mechanics all the manual arts belong, from which mechanics took its name. But as artificers do not work with perfect accuracy, it comes to pass that mechanics is so distinguished from geometry that which is perfectly accurate is called geometrical; what is less so, is called mechanical. However, the errors are not in the art, but in the artificer. He that works with less accuracy is an imperfect mechanic; and if any could work with perfect accuracy, he would be the most perfect mechanic of all, for the description of right lines and circles, upon which geometry is founded, belongs to mechanics. Geometry does not teach us to draw these lines, but requires them to be drawn, for it requires that the learner should first be taught to describe these accurately before he enters upon geometry, then it shows how by these operations problems may be solved. To describe right lines and circles are problems, but not geometrical problems. The solution of these problems is required from mechanics, and by Geometry the use of them, when so solved, is shown; and it is the glory of Geometry that from those few principles, brought from without, it is able to produce so many things. Therefore geometry is founded in mechanical practice, and is nothing but that part of universal mechanics which accurately proposes and demonstrates the art of measuring. But since the manual arts are chiefly employed in the moving of bodies, it happens that Geometry is commonly referred to their magnitude, and mechanics to their motion. In this sense rational mechanics will be the science of motions resulting from any forces whatsoever, and of the forces required to produce any motions, accurately proposed and demonstrated. This part of mechanics, as far as it extended to the five powers which relate to manual arts, was cultivated by the ancients, who considered gravity (it not being a manual power) no otherwise than in moving weights by those powers. But I consider philosophy rather than arts and write not concerning manual but natural powers, and consider chiefly those things which relate to gravity, levity, elastic force, the resistance of fluids. and the like forces, whether attractive or impulsive, and therefore I offer this work as the mathematical principles of philosophy, for the whole burden of philosophy deems to consist in this—from the phenomena of motions to investigate the forces of nature, and then from these forces to demonstrate the other phenomena; and to this end the general propositions in the first and second Books are directed. In the third Book I give an example of this in the explication of the System of the World, for by the propositions mathematically demonstrated in the former Books, in the third I derive from the celestial phenomena the forces of gravity with which bodies tend to the sun and the several planets. Then from these forces, by other propositions which are also mathematical, I deduce the motions of the planets, the comets, the moon, and the sea. I wish we could derive the rest of the phenomena of Nature by the same kind of reasoning from mechanical principles, for I am induced by many reasons to suspect that they may all depend upon certain forces by which the particles of bodies, by some causes hitherto unknown, are either mutually impelled towards one another, and cohere in regular figures, or are repelled and recede from one another. These forces being unknown, philosophers have hitherto attempted the search of Nature in vain; but I hope the principles here laid down will afford some light either to this or some truer method of philosophy.

In the publication of this work the most acute and universally learned Mr. *Edmund Halley* not only assisted me in correcting the errors of the press and preparing the geometrical figures, but it was through his solicitations that it came to be published; for when he had obtained of me my demonstrations of the figure of the celestial orbits, he continually pressed me to communicate the same to the *Royal Society*, who afterwards, by their kind encouragement and entreaties, engaged me to think of publishing them. But after I had begun to consider the inequalities of the lunar motions, and had entered upon some other things relating to the laws and measures of gravity and other forces; and the figures that would be described by bodies attracted according to given laws; and the motion of several bodies moving among themselves; the motion of bodies in resisting mediums; the forces, densities, and motions, of mediums; the orbits of the comets, and such like, I deferred that publication till I had made a search into those matters, and could put forth the whole together. What relates to the lunar motions (being imperfect), I have put all together in the corollaries of Prop. LXVI, to avoid being obliged to propose and distinctly demonstrate the several things there contained in a method more prolix than the subject deserved and interrupt the series of the other propositions. Some things, found out after the rest, I chose to insert in places less suitable, rather than change the number of the propositions and the citations. I heartily beg that what I have here done may be read with forbearance; and that my labors in a subject so difficult may be examined, not so much with the view to censure as to remedy their defects.

Is. NEWTON
Cambridge, Trinity College, May 8. 1686[2]

SCHOLIUM[3]

Hitherto I have laid down the definitions of such words as are less known and explained the sense in which I would have them to be understood in the following discourse. I do not define time, space, place, and motion, as being well known to all. Only I must observe, that the common people conceive those quantities under no other notions but from the relation they bear to sensible objects. And thence arise certain prejudices, for the removing of which it will be convenient to distinguish them into absolute and relative, true and apparent, mathematical and common.

I.   Absolute, true, and mathematical time, of itself, and from its own nature, flows equably without relation to anything external, and by another name is called duration: relative, apparent, and common time, is some sensible and external (whether accurate or unequable) measure of duration by the means of motion, which is commonly used instead of true time; such as an hour, a day, a month, a year.

II. Absolute space, in its own nature, without relation to anything external, remains always similar and immovable. Relative space is some movable dimension or measure of the absolute spaces; which our senses determine by its position to bodies; and which is commonly taken for immovable space; such is the dimension of a subterraneous, an aerial, or celestial space, determined by its position in

---

[2]   Appendix, Note 3.

[3]   Appendix, Note 13.

respect of the earth. Absolute and relative space are the same in figure and magnitude; but they do not remain always numerically the same. For if the earth, for instance, moves, a space of our air, which relatively and in respect of the earth remains always the same, will at one time be one part of the absolute space into which the air passes; at another time it will be another part of the same, and so, absolutely understood, it will be continually changed.

III. Place is a part of space which a body takes up, and is according to the space, either absolute or relative. I say, a part of space; not the situation, nor the external surface of the body. For the places of equal solids are always equal; but their surfaces, by reason of their dissimilar figures, are often unequal. Positions properly have no quantity, nor are they so much the places themselves, as the properties of places. The motion of the whole is the same with the sum of the motions of the parts; that is, the translation of the whole, out of its place, is the same thing with the sum of the translations of the parts out of their places; and therefore the place of the whole is the same as the sum of the places of the parts, and for that reason, it is internal, and in the whole body.

IV. Absolute motion is the translation of a body from one absolute place into another; and relative motion, the translation from one relative place into another. Thus in a ship under sail, the relative place of a body is that part of the ship which the body possesses; or that part of the cavity which the body fills, and which therefore moves together with the ship: and relative rest is the continuance of the body in the same part of the ship, or of its cavity. But real, absolute rest, is the continuance of the body in the same part of that immovable space, in which the ship itself, its cavity, and all that it contains, is moved. Wherefore, if the earth is really at rest, the body, which relatively rests in the ship, will really and absolutely move with the same velocity which the ship has on the earth. But if the earth also moves, the true and absolute motion of the body will arise, partly from the true motion of the earth, in immovable space partly from the relative motion of the ship on the earth; and if the body moves also relatively in the ship, its true motion will arise, partly from the true motion of the earth, in immovable space, and partly from the relative motions as well of the ship on the earth, as of the body in the ship; and from these relative motions will arise the relative motion of the body on the earth. As if that part of the earth, where the ship is, was truly moved towards the east, with a velocity of 10,010 parts, while the ship itself, with a fresh gale, and full sails, is carried towards the west, with a velocity expressed by 10 of those parts; but a sailor walks in the ship towards the east, with 1 part of the said velocity, then the sailor will be moved truly in immovable space towards the east, with a velocity of 10,001 parts, and relatively on the earth towards the west, with a velocity of 9 of those parts.

Absolute time, in astronomy, is distinguished from relative, by the equation or correction of the apparent time. For the natural days are truly unequal, though they are commonly considered as equal, and used for a measure of time; astronomers correct this inequality that they may measure the celestial motions by a more accurate time. It may be, that there is no such thing as an equable motion, whereby time may be accurately measured. All motions may be accelerated and retarded, but the flowing of absolute time is not liable to any change. The duration of perseverance of the existence of things remains the same, whether the motions are swift or slow, or none at all: and therefore this duration ought to be distinguished from what are only sensible measures thereof; and from which we deduce it, by means of the astronomical equation. The necessity of this equation, for determining the times of a phenomenon, is evinced as well from the experiments of the pendulum clock, as by eclipses of the satellites of Jupiter.

As the order of the parts of time is immutable, so also is the order of the parts of space. Suppose those parts to be moved out of their places, and they will be moved (if the expression may be allowed) out of themselves. For times and spaces are, as it were, the places as well of themselves as of all other things. All things are placed in time as to order of succession; and in space as to order of situation. It is from their essence or nature that they are places; and that the primary places of things should be movable, is absurd. These are therefore the absolute places; and translations out of those places, are the only absolute motions.

But because the parts of space cannot be seen, or distinguished from one another by our senses, therefore in their stead we use sensible measures of them. For from the positions and distances of things from any body considered as immovable, we define all places; and then with respect to such places, we estimate all motions, considering bodies as transferred from some of those places into others. And so, instead of absolute places and motions, we use relative ones; and that without any inconvenience in common affairs; but in philosophical disquisitions, we ought to abstract from our senses, and consider things themselves, distinct from what are only sensible measures of them. For it may be that there is no body really at rest, to which the places and motions of others may be referred.

But we may distinguish rest and motion, absolute and relative, one from the other by their properties, causes, and effects. It is a property of rest, that bodies really at rest do rest in respect to one another. And therefore as it is possible, that in the remote regions of the fixed stars, or perhaps far beyond them, there may be some body absolutely at rest; but impossible to know, from the position of bodies to one another in our regions, whether any of these do keep the same position to that remote body, it follows that absolute rest cannot be determined from the position of bodies in our regions.

It is a property of motion, that the parts, which retain given positions to their wholes, do partake of the motions of those wholes. For all the parts of revolving bodies endeavor to recede from the axis of motion; and the impetus of bodies moving forwards arises from the joint impetus of all the parts. Therefore, if surrounding bodies are moved, those that are relatively at rest within them will partake of their motion. Upon which account, the true and absolute motion of a body cannot be determined by the translation of it from those which only seem to rest; for the external bodies ought not only to appear at rest, but to be really at rest. For otherwise, all included bodies, besides their translation from near the surrounding ones, partake likewise of their true motions; and though that translation were not made, they would not be really at rest, but only seem to be so. For the surrounding bodies stand in the like relation to the surrounded as the exterior part of a whole does to the interior, or as the shell does to the kernel; but if the shell moves, the kernel will also move, as being part of the whole, without any removal from near the shell.

A property, near akin to the preceding, is this, that if a place is moved, whatever is placed therein moves along with it; and therefore a body, which is moved from a place in motion, partakes also of the motion of its place. Upon which account, all motions, from places in motion, are no other than parts of entire and absolute motions; and every entire motion is composed of the motion of the body out of its first place, and the motion of this place out of its place, and so on, until we come to some immovable place, as in the before-mentioned example of the sailor. Wherefore, entire and absolute motions can be no otherwise determined than by immovable places; and for that reason I did before refer those absolute motions to immovable places, but relative ones to movable places. Now no other places are immovable but those that, from infinity to infinity, do all retain the same given position one to another; and upon this account must ever remain unmoved; and do thereby constitute immovable space.

The causes by which true and relative motions are distinguished, one from the other, are the forces impressed upon bodies to generate motion. True motion is neither generated nor altered, but by some force impressed upon the body moved; but relative motion may be generated or altered without any force impressed upon the body. For it is sufficient only to impress some force on other bodies with which the former is compared, that by their giving way, that relation may be changed, in which the relative rest or motion of this other body did consist. Again, true motion suffers always some change from any force impressed upon the moving body; but relative motion does not necessarily undergo any change by such forces. For if the same forces are likewise impressed on those other bodies with which the comparison is made, that the relative position may be preserved, then that condition will be preserved in which the relative motion consists. And therefore any relative motion may be changed when the true motion remains unaltered, and the relative may be preserved when the true suffers some change. Thus, true motion by no means consists in such relations.

The effects which distinguish absolute from relative motion are, the forces of receding from the axis of circular motion. For there are no such forces in a circular motion purely relative, but in a true and absolute circular motion, they are greater or less, according to the quantity of the motion. If a vessel, hung by a long cord, is so often turned about that the cord is strongly twisted, then filled with water, and held at rest together with the water; thereupon, by the sudden action of another force, it is whirled about the contrary way, and while the cord is untwisting itself, the vessel continues for some time in this motion; the surface of the water will at first be plain, as before the vessel began to move, but after that, the vessel, by gradually communicating its motion to the water, will make it begin sensibly to revolve, and recede by little and little from the middle, and ascend to the sides of the vessel, forming itself into a concave figure (as I have experienced), and the swifter the motion becomes, the higher will the water rise, till at last, performing its revolutions in the same times with the vessel it becomes relatively at rest in it. This ascent of the water shows its endeavor to recede from the axis of its motion; and the true and absolute circular motion of the water, which is here directly contrary to the relative, becomes known, and may be measured by this endeavor. At first, when the relative motion of the water in the vessel was greatest, it produced no endeavor to recede from the axis; the water showed no tendency to the circumference, nor any ascent towards the sides of the vessel, but remained of a plain surface, and therefore its true circular motion had not yet begun. But afterwards, when the relative motion of the water had decreased, the ascent thereof towards the sides of the vessel proved its endeavor to recede from the axis; and this endeavor showed the real circular motion of the water continually increasing, till it had acquired its greatest quantity, when the water rested relatively in the vessel. And therefore this endeavor does not depend upon any translation of the water in respect of the ambient bodies, nor can true circular motion be defined by such translation. There is only one real circular motion of any one revolving body, corresponding to only one power of endeavoring to recede from its axis of motion, as its proper and adequate effect; but relative motions, in one and the same body, are innumerable, according to the various relations it bears to external bodies, and, like other relations, are altogether destitute of any real effect, any otherwise than they may perhaps partake of that one only true motion. And therefore in their system who suppose that our heavens, revolving below the sphere of the fixed stars, carry the planets along with them; the several parts of those heavens, and the planets, which are indeed relatively at rest in their heavens, do yet really move. For they change their position one to another (which never happens to bodies truly at rest), and being carried together with their heavens, partake of their motions, and as parts of revolving wholes, endeavor to recede from the axis of their motions.

Wherefore relative quantities are not the quantities themselves, whose names they bear, but those sensible measures of them (either accurate or inaccurate), which are commonly used instead of the measured quantities themselves. And if the meaning of words is to be determined by their use, then by the names time, space, place, and motion, their [sensible] measures are properly to be understood; and the expression will be unusual, and purely mathematical, if the measured quantities themselves are meant. On this account, those violate the accuracy of language, which ought to be kept precise, who interpret these words for the measured quantities. Nor do those less defile the purity of mathematical and philosophical truths, who confound real quantities with their relations and sensible measures.

It is indeed a matter of great difficulty to discover, and effectually to distinguish, the true motions of particular bodies from the apparent; because the parts of that immovable space, in which those motions are performed, do by no means come under the observation of our senses. Yet the thing is not altogether desperate; for we have some arguments to guide us, partly from the apparent motions, which are the differences of the true motions; partly from the forces, which are the causes and effects of the true motions. For instance, if two globes, kept at a given distance one from the other by means of a cord that connects them, were revolved about their common centre of gravity, we might, from the tension of the cord, discover the endeavor of the globes to recede from the axis of their motion, and from thence we might compute the quantity of their circular motions. And then if any equal forces should be impressed at once on the alternate faces of the globes to augment or diminish their circular motions, from the increase or decrease of the tension of the cord, we might infer the increment or decrement of their motions; and thence would be found on what faces those forces ought to be impressed, that the motions of the globes might be most augmented; that is, we might discover their hindmost faces, or those which, in the circular motion, do follow. But the faces which follow being known, and consequently the opposite ones that precede, we should likewise know the determination of their motions. And thus we might find both the quantity and the determination of this circular motion, even in an immense vacuum, where there was nothing external or sensible with which the globes could be compared. But now, if in that space some remote bodies were placed that kept always a given position one to another, as the fixed stars do in our regions, we could not indeed determine from the relative translation of the globes among those bodies, whether the motion did belong to the globes or to the bodies. But if we observed the cord and found that its tension was that very tension which the motions of the globes required, we might conclude the motion to be in the globes, and the bodies to be at rest; and then, lastly, from the translation of the globes among the bodies, we should find the determination of their motions. But how we are to obtain the true motions from their causes, effects, and apparent differences, and the converse, shall be explained more at large in the following treatise. For to this end it was that I composed it.

AXIOMS, OR
LAWS OF MOTION[4]

LAW I

Every body continues in its state of rest, or of uniform motion in a right line, unless it is compelled to change that state by forces impressed upon it.

Projectiles continue in their motions, so far as they are not retarded by the resistance of the air, or impelled downwards by the force of gravity. A top, whose parts by their cohesion are continually drawn aside from rectilinear motions, does not cease its rotation, otherwise than as it is retarded by the air. The greater bodies of the planets and comets, meeting with less resistance in freer spaces, preserve their motions both progressive and circular for a much longer time.

LAW II[5]

The change of motion is proportional to the motive force impressed; and is made in the direction of the right line in which that force is impressed.

---

[4]   Appendix, Note 14.

[5]   Appendix, Note 15.

If any force generates a motion, a double force will generate double the motion, a triple force triple the motion, whether that force be impressed altogether and at once, or gradually and successively. And this motion (being always directed the same way with the generating force), if the body moved before, is added to or subtracted from the former motion, according as they directly conspire with or are directly contrary to each other; or obliquely joined, when they are oblique, so as to produce a new motion compounded from the determination of both.

LAW III

To every action there is always opposed an equal reaction: or, the mutual actions of two bodies upon each other are always equal, and directed to contrary parts.

Whatever draws or presses another is as much drawn or pressed by the other. If you press a stone with your finger, the finger is also pressed by the stone. If a horse draws a stone tied to a rope, the horse (if I may so say) will be equally drawn back towards the stone; for the distended rope, by the same endeavor to relax or unbend itself, will draw the horse as much towards the stone as it does the stone towards the horse, and will obstruct the progress of the one as much as it advances that of the other. If a body impinge upon another, and by its force change the motion of the other, that body also (because of the equality of the mutual pressure) will undergo an equal change, in its own motion, towards the contrary part. The changes made by these actions are equal, not in the velocities but in the motions of bodies; that is to say, if the bodies are not hindered by any other impediments. For, because the motions are equally changed, the changes of the velocities made towards contrary parts are inversely proportional to the bodies. This law takes place also in attractions, as will be proved in the next Scholium . . . .

RULES OF REASONING
IN PHILOSOPHY

RULE I

We are to admit no more causes of natural things than such as are both true and sufficient to explain their appearances.

To this purpose the philosophers say that Nature does nothing in vain, and more is in vain when less will serve; for Nature is pleased with simplicity, and affects not the pomp of superfluous causes.

RULE II

Therefore to the same natural effects we must, as far as possible, assign the same causes.

As to respiration in a man and in a beast; the descent of stones in Europe and in America; the light of our culinary fire and of the sun; the reflection of light in the earth, and in the planets.

RULE III

The qualities of bodies, which admit neither intensification nor remission of degrees, and which are found to belong to all bodies within the reach of our experiments, are to be esteemed the universal qualities of all bodies whatsoever.

For since the qualities of bodies are only known to us by experiments, we are to hold for universal all such as universally agree with experiments; and such as are not liable to diminution can never be quite taken away. We are certainly not to relinquish the evidence of experiments for the sake of dreams and vain fictions of our own devising; nor are we to recede from the analogy of Nature, which is wont to be simple, and always consonant to itself. We no other way know the extension of bodies than by our senses, nor do these reach it in all bodies; but because we perceive extension in all that are sensible, therefore describe it universally to all others also. That abundance of bodies are hard, we learn by experience; and because the hardness of the whole arises from the hardness of the parts, we therefore justly infer the hardness of the undivided particles not only of the bodies we feel but of all others. That all bodies are impenetrable, we gather not from reason, but from sensation. The bodies which we handle we find impenetrable, and thence conclude impenetrability to be an universal property of all bodies whatsoever. That all bodies are movable, and endowed with certain powers (which we call the inertia) of persevering in their motion, or in their rest, we only infer from the like properties observed in the bodies which we have seen. The extension, hardness, impenetrability, mobility, and inertia of the whole, result from the extension, hardness impenetrability, mobility, and inertia of the parts; and hence we conclude the least particles of all bodies to be also all extended and hard and impenetrable, and movable, and endowed with their proper inertia. And this is the foundation of all philosophy. Moreover, that the divided but contiguous particles of bodies may be separated from one another, is matter of observation, and, in the particles that remain undivided, our minds are able to distinguish yet lesser parts, as is mathematically demonstrated. But whether the parts so distinguished, and not yet divided, may by the powers of Nature, be actually divided and separated from one another, we cannot certainly determine. Yet, had we the proof of but one experiment that any undivided particle, in breaking a hard and solid body, suffered a division, we might by virtue of this rule conclude that the undivided as well as the divided particles may be divided and actually separated to infinity.

Lastly, if it universally appears, by experiments and astronomical observations, that all bodies about the earth gravitate towards the earth, and that in proportion to the quantity of matter which they severally contain; that the moon likewise, according to the quantity of its matter, gravitates towards the earth; that, on the other hand, our sea gravitates towards the moon; and all the planets one towards another, and the comets in like manner towards the sun; we must, in consequence of this rule, universally allow that all bodies whatsoever are endowed with a principle of mutual gravitation. For the argument from the appearances concludes with more force for the universal gravitation of all bodies than for their impenetrability; of which, among those in the celestial regions, we have no experiments, nor any manner of observation. Not that I affirm gravity to be essential to bodies: by their *vis insita* I mean nothing but their inertia. This is immutable. Their gravity is diminished as they recede from the earth.

RULE IV

In experimental philosophy we are to look upon propositions inferred by general inductions from phenomena as accurately or very nearly true, notwithstanding any contrary hypotheses that may be imagined, till such time as other phenomena occur, by which they may either be made more accurate, or liable to exceptions.

This rule we must follow, that the argument of induction may not be evaded by hypotheses.

Michael Faraday

1791-1864

The Chemical History of a Candle

1861

The "Faraday effect" in physics, "Faraday's law of induction," and "Faraday's laws of electrolysis" are all named for discoveries by the English physicist and chemist, Michael Faraday (1791-1867). Faraday discovered a number of new organic compounds and wrote a manual of practical chemistry, but more importantly, he was the first to develop an electric current from a magnetic field. He invented the dynamo and the electric motor. To Faraday's work we owe the terms *electrode, cathode, ion, anion, cation, ionization, electrolyte,* and *electrolysis.*

Born to a poor family in Surrey, England, Faraday learned to read, write, and cipher in Sunday School. When apprenticed to a book dealer and bookbinder at age 14, he began to read the many books on chemistry and science available to him and to discuss his ideas in the City Philosophical Society, an organization of young men and women which was devoted to self-improvement. Its members met every week to hear lectures on scientific topics and to discuss scientific matters. Toward the end of his apprenticeship, he heard 4 lectures given Sir Humphry Davy. Faraday was so impressed that when he was free to seek a position, he applied to Davy at the Royal

Institution's labs, and before long was appointed to a position as a Chemical Assistant and then as one of Davy's laboratory assistants.

Ten years later, he became superintendent of the Royal Institution and, in 1825, succeeded Davy as director of the laboratory where he became a Fullerian professor of chemistry in 1833. He was elected a member of the Royal Society but declined many other honors, including a knighthood. In 1858 he retired to a house at Hampton Court presented to him by Queen Victoria.

The Chemical History of a Candle was a series of six lectures presented to young people at the Royal Institution, London 1860. "I . . . bring before you, in the course of these lectures, the Chemical History of a Candle. There is not a law under which any part of this universe is governed which does not come into play and is touched upon in these phenomena. . . . There is no more open door by which you can enter into the study of natural philosophy than by considering the physical phenomena of a candle."

Sources:

"Faraday, Michael." 2007. In *Encyclopædia Britannica.* Retrieved August 23, 2007, from Encyclopædia Britannica Online: *http://www.britannica.com/eb/article-9109756*

"Heritage: Faraday Page." 2007. The Royal Institution of Great Britain. Retrieved August 23, 2007, from http://www.rigb. org/rimain/heritage/faradaypage.jsp

"Michael Faraday." 2002. School of Mathematics and Statistics, University of St. Andrews, Scotland. Retrieved February 22, 2007, from http://www.gap-system.org/~history/ Biographies/Faraday.html.

Weisstein, Eric. 2007. "Faraday, Michael." scienceworld. wolfram.com. Retrieved February 22, 2007, from http:// scienceworld.wolfram.com/biography/Faraday.html.

Selection From:

Faraday, Michael. 1861. *The Chemical History of the Candle.* Lectures I, II, III. In *The Harvard Classics, Selected Paper—Physics, Chemistry, Astronomy, Geology.* Volume 30. Ed. Charles W. Eliot. New York: P.F. Collier. 1910. 89-146.

LECTURE I

A CANDLE: THE FLAME—ITS SOURCE— STRUCTURE—MOBILITY—BRIGHTNESS

I PURPOSE, in return for the honor you do us by coming to see what are our proceedings here, to bring before you, in the course of these lectures, the Chemical History of a Candle. I have taken this subject on a former occasion, and, were it left to my own will, I should prefer to repeat it almost every year, so abundant is the interest that attaches itself to the subject, so wonderful are the varieties of outlet which it offers into the various departments of philosophy. There is not a law under which any part of this universe is governed which does not come into play and is touched upon in these phenomena. There is no better, there is no more open door by which you can enter into the study of natural philosophy than by considering the physical phenomena of a candle. I trust, therefore, I shall not disappoint you in choosing this for my subject rather than any newer topic, which could not be better, were it even so good.

And before proceeding, let me say this also: that, though our subject be so great, and our intention that of treating it honestly, seriously, and philosophically, yet I mean to pass away from all those who are seniors among us. I claim the privilege of speaking to juveniles as a juvenile myself. I have done so on former occasions, and, if you please, I shall do so again. And, though I stand here with the knowledge of having the words I utter given to the world, yet that shall not deter me from speaking in the same familiar way to those whom I esteem nearest to me on this occasion.

And now, my boys and girls, I must first tell you of what candles are made. Some are great curiosities. I have here some bits of timber, branches of trees, particularly famous for their burning. And here you see a piece of that very curious substance, taken out of some of the bogs in Ireland, called *candle-wood*; a hard, strong, excellent wood, evidently fitted for good work as a register of force, and yet, withal, burning so well that where it is found they make splinters of it, and torches, since it burns like a candle, and gives a very good light indeed. And in this wood we have one of the most beautiful illustrations of the general nature of a candle that I can possibly give. The fuel provided, the means of bringing that fuel to the place of chemical action, the regular and gradual supply of air to that place of action—heat and light—all produced by a little piece of wood of this kind, forming, in fact, a natural candle.

But we must speak of candles as they are in commerce. Here are a couple of candles commonly called dips. They are made of lengths of cotton cut off, hung up by a loop, dipped into melted tallow, taken out again and cooled, then redipped, until there is an accumulation of tallow round the cotton. In order that you may have an idea of the various characters of these candles, you see these which I hold in my hand—they are very small and very curious. They are, or were, the candles used by the miners in coal mines. In olden times the miner had to find his own candles, and it was supposed that a small candle would not so soon set fire to the fire-damp in the coal mines as a large one; and for that reason, as well as for economy's sake, he had candles made of this sort—20, 30, 40, or 60 to the pound. They have been replaced since then by the steel-mill, and then by the Davy lamp, and other safety lamps of various kinds. I have here a candle that was taken out of the *Royal George*,[1] it is said, by Colonel Pasley. It has been sunk in the sea for many years, subject to the action of salt water. It shows you how well candles may be preserved; for, though it is cracked about and broken a great deal, yet when lighted it goes on burning regularly, and the tallow resumes its natural condition as soon as it is fused.

Mr. Field, of Lambeth, has supplied me abundantly with beautiful illustrations of the candle and its materials; I shall therefore now refer to them. And, first, there is the suet—that fat of the ox—Russian tallow, I believe, employed in the manufacture of these dips, which Gay-Lussac, or some one who intrusted him with his knowledge, converted into that beautiful substance, stearin, which you see lying beside it. A candle, you know, is not now a greasy thing like an ordinary tallow candle, but a clean thing, and you may almost scrape off and pulverize the drops which fall from it without soiling any thing. This is the process he adopted:[2]

---

[1]   The *Royal George* sunk at Spithead on the 29th of August, 1782. Colonel Pasley commenced operations for the removal of the wreck by the explosion of gunpowder, in August, 1839. The candle which Professor Faraday exhibited must therefore have been exposed to the action of salt water for upward of fifty-seven years.

[2]   The fat or tallow consists of a chemical combination of fatty acids with glycerin. The lime unites with the palmitic, oleic, and stearic acids, and separates the glycerin. After washing, the insoluble lime soap is decomposed with hot dilute sulphuric acid. The melted fatty acids thus rise as an oil to the surface, when they are decanted. They are again washed and cast into thin plates, which, when cold, are placed between layers of cocoanut matting and submitted to intense hydraulic pressure. In this way the soft oleic acid is squeezed out, while the hard palmitic and stearic acids remain. These are farther purified by pressure at a higher temperature and washing in warm dilute sulphuric acid, when they are ready to be made in to candles. These acids are harder and whiter than the fats from which they were obtained, while at the same time they are cleaner and more combustible.

The fat or tallow is first boiled with quick-lime, and made into a soap, and then the soap is decomposed by sulphuric acid, while a quantity of glycerin is produced at the same time. Glycerin—absolutely a sugar, or a substance similar to sugar—comes out of the tallow in this chemical change. The oil is then pressed out of it; and you see here this series of pressed cakes, showing how beautifully the impurities are carried out by the oil part as the pressure goes on increasing, and at last you have left that substance, which is melted, and cast into candles as here represented. The candle I have in my hand is a stearin candle, made of stearin from tallow in the way I have told you. Then here is a sperm candle, which comes from the purified oil of the spermaceti whale. Here, also, are yellow bee's-wax and refined bee's-wax, from which candles are made. Here, too, is that curious substance called paraffine, and some paraffine candles, made of paraffine obtained from the bogs of Ireland. I have here also a substance brought from Japan since we have forced an entrance into that out-of-the-way place—a sort of wax which a kind friend has sent me, and which forms a new material for the manufacture of candles.

And how are these candles made? I have told you about dips, and I will show you how moulds are made. Let us imagine any of these candles to be made of materials which can be cast. "Cast!" you say. "Why, a candle is a thing that melts, and surely if you can melt it you can cast it." Not so. It is wonderful, in the progress of manufacture, and in the consideration of the means best fitted to produce the required result, how things turn up which one would not expect beforehand. Candles can not always be cast. A wax candle can never be cast. It is made by a particular process which I can illustrate in a minute or two, but I must not spend much time on it. Wax is a thing which, burning so well, and melting so easily in a candle, can not be cast. However, let us take a material that can be cast. Here is a frame, with a number of moulds fastened in it. The first thing to be done is to put a wick through them. Here is one—a plaited wick, which does not require snuffing[3]—supported by a little wire. It goes to the bottom, where it is pegged in; the little peg holding the cotton tight, and stopping the aperture so that nothing but fluid shall run out. At the upper part there is a little bar placed across, which stretches the cotton and holds it in the mould. The tallow is then melted, and the moulds are filled. After a certain time, when the moulds are cool, the excess of tallow is poured off at one corner, and then cleaned off altogether, and the ends of the wick cut away. The candles alone then remain in the mould, and you have only to upset them, as I am doing, when out they tumble, for the candles are made in the form of cones, being narrower at the top than at the bottom; so that, what with their form and their own shrinking, they only need a little shaking, and out they fall. In the same way are made these candles of stearin and of paraffine. It is a curious thing to see how wax candles are made. A lot of cottons are hung upon frames, as you see here, and covered with metal tags at the ends to keep the wax from covering the cotton in those places. These are carried to a heater, where the wax is melted. As you see, the frames can turn round; and, as they turn, a man takes a vessel of wax and pours it first down one, and then the next, and the next, and so on. When he has gone once round, if it is sufficiently cool, he gives the first a second coat, and so on until they are all of the required thickness. When they have been thus clothed, or fed, or made up to that thickness, they are taken off and placed elsewhere. I have here, by the kindness of Mr. Field, several specimens of these candles. Here is one only half finished. They are then taken down and well rolled upon a fine stone slab, and the conical top is moulded by properly shaped tubes, and the bottoms cut off and trimmed. This is done so beautifully that they can make candles in this way weighing exactly four or six to the pound, or any number they please.

---

[3]   A little borax or phosphorus added in order to make the ash fusible.

We must not, however, take up more time about the mere manufacture, but go a little farther into the matter. I have not yet referred you to luxuries in candles (for there is such a thing as luxury in candles). See how beautifully these are colored; you see here mauve, magenta, and all the chemical colors recently introduced, applied to candles. You observe, also, different forms employed. Here is a fluted pillar most beautifully shaped; and I have also here some candles sent me by Mr. Pearsall, which are ornamented with designs upon them, so that, as they burn, you have, as it were, a glowing sun above, and bouquet of flowers beneath. All, however, that is fine and beautiful is not useful. These fluted candles, pretty as they are, are bad candles; they are bad because of their external shape. Nevertheless, I show you these specimens, sent to me from kind friends on all sides, that you may see what is done and what may be done in this or that direction; although, as I have said, when we come to these refinements, we are obliged to sacrifice a little in utility.

Now as to the light of the candle. We will light one or two, and set them at work in the performance of their proper functions. You observe a candles is a very different thing from a lamp. With a lamp you take a little oil, fill your vessel, put in a little moss or some cotton prepared by artificial means, and then light the top of the wick. When the flame runs down the cotton to the oil, it gets extinguished, but it goes on burning in the part above. Now I have no doubt you will ask how it is that the oil which will not burn of itself gets up to the top of the cotton, where it will burn. We shall presently examine that; but there is a much more wonderful thing about the burning of a candle than this. You have here a solid substance with no vessel to contain it; and how it is that this solid substance can get up to the place where the flame is? How is it that this solid get there, it not being a fluid? or, when it is made a fluid, then how is it that it keeps together? This is a wonderful thing about a candle.

We have here a good deal of wind, which will help us in some of our illustrations, but tease us in others; for the sake, therefore, of a little regularity, and to simplify the matter, I shall make a quiet flame, for who can study a subject when there are difficulties in the way not belonging to it? Here is a clever invention of some costermonger or street-stander in the market-place for the shading of their candles on Saturday nights, when they are selling their greens, or potatoes, or fish. I have very often admired it. They put a lamp-glass round the candle, supported on a kind of gallery, which clasps it, and it can be slipped up and down as required. By the use of this lamp-glass, employed in the same way, you have a steady flame, which you can look at, and carefully examine, as I hope you will do, at home.

You see then, in the first instance, that a beautiful cup is formed. As the air comes to the candle, it moves upward by the force of the current which the heat of the candle produces, and it so cools all the sides of the wax, tallow, or fuel as to keep the edge much cooler than the part within; the part within melts by the flame that runs down the wick as far as it can go before it is extinguished, but the part on the outside does not melt. If I made a current in one direction, my cup would be lop-sided, and the fluid would consequently run over; for the same force of gravity which holds worlds together holds this fluid in a horizontal position, and if the cup be not horizontal, of course the fluid will run away in guttering. You see, therefore, that the cup is formed by this beautifully regular ascending current of air playing upon all sides, which keeps the exterior of the candle cool. No fuel would serve for a candle which has not the property of giving this cup, except such fuel as the Irish bog-wood, where the material itself is like a sponge and holds its own fuel. You see now why you would have had such a bad result if you were to burn these beautiful candles that I have shown you, which are irregular, intermittent in their shape, and can not, therefore, have that nicely-formed edge to the cup which is the great beauty in a candle. I hope you will now see that the perfection of a process—that is, its utility—is the better point of beauty about it. It is not the best looking thing, but the best acting thing, which is the most advantageous to us. This good-looking candle is a bad-burning one. There will be a guttering round about it because of the irregularity of the stream of air and the badness of the cup which is formed thereby. You may see some pretty examples (and I trust you will notice these instances) of the action of the ascending current when you have a little gutter run down the side of a candle, making it thicker there than it is elsewhere. As the candle goes on burning, that keeps its place and forms a little pillar sticking up by the side, because, as it rises higher above the rest of the wax or fuel, the air gets better round it, and it is more cooled and better able to resist the action of the heat at a little distance. Now the greatest mistakes and faults with regard to candles, as in many other things, often bring with them instruction which we should not receive if they had not occurred. We come here to be philosophers, and I hope you will always remember that whenever a result happens, especially if it be new, you should say, "What is the cause? Why does it occur?" and you will, in the course of time, find out the reason. Then there is another point about these candles which will answer a question—that is, as to the way in which this fluid gets out of the cup, up the wick, and into the place of combustion. You know that the flames on these burning wicks in candles made of bees'-wax, stearin, or spermaceti, do not run down to the wax or other matter, and melt it all away, but keep to their own right place. They are fenced off from the fluid below, and do not encroach on the cup at the sides. I can not imagine a more beautiful example than the condition of adjustment under which a candle makes one part subserve to the other to the very end of its action. A combustible thing like that, burning away gradually, never being intruded upon by the flame, is a very beautiful sight, especially when you come to learn what a vigorous thing flame is—what power it has of destroying the wax itself when it gets hold of it, and of disturbing its proper form if it come only too near.

But how does the flame get hold of the fuel? There is a beautiful point about that—*capillary attraction*.[4] "Capillary attraction!" you say—"the attraction of hairs." "Capillary attraction!" you say—"the attraction of hairs." Well, never mind the name; it was given in old times, before we had a good understanding of what the real power was. It is by what is called capillary attraction that the fuel is conveyed to the part where combustion goes on, and is deposited there, not in a careless way, but very beautifully in the very midst of the centre of action, which takes place around it. Now I am going to give you one or two instances of capillary attraction. It is that kind of action or attraction which makes two things that do not dissolve in each other still hold together. When you wash your hands, you wet them thoroughly; you take a little soap to make the adhesion better, and you find your hands remain wet. This is by that kind of attraction of which I am about to speak. And, what is more, if your hands are not soiled (as they almost always are by the usages of life), if you put your finger into a little warm water, the water will creep a little way up the finger, though you may not stop to examine it. I have here a substance which is rather porous—a column of salt—and I will pour into the plate at the bottom, not water, as it appears, but a saturated solution of salt which can not absorb more, so that the action which you see will not be due to its dissolving any thing. We may consider the plate to be the candle, and the salt the wick, and this solution the melted tallow. (I have colored the fluid, that you may see the action better.) You

---

[4]   Capillary attraction or repulsion is the cause which determines the ascent or descent of a fluid in a capillary tube. If a piece of thermometer tubing, open at each end, be plunged into water, the latter will instantly rise in the tube considerably above its external level. If, on the other hand, the tube be plunged into mercury, a repulsion instead of attraction will be exhibited, and the level of the mercury will be lower in the tube than it is outside.

observe that, now I pour in the fluid, it rises and gradually creeps up the salt higher and higher (Fig. 55); and provided the column does not tumble over, it will go to the top. If this blue solution were combustible, and we were to place a wick at the top of the salt, it would burn as it entered into the wick. It is a most curious thing to see this kind of action taking place, and to observe how singular some of

FIG. 55

the circumstances are about it. When you wash your hands, you take a towel to wipe off the water; and it is by that kind of wetting, or that kind of attraction which makes the towel become wet with water, that the wick is made wet with tallow. I have known some careless boys and girls (indeed, I have known it happen to careful people as well) who, having washed their hands and wiped them with a towel, have thrown the towel over the side of the basin, and before long it has drawn all the water out of the basin and conveyed it to the floor, because it happened to be thrown over the side in such a way as to serve the purpose of a siphon.[5] That you may the better see the way in which the substances act one upon another, I have here a vessel made of wire gauze filled with water, and you may compare it in its action to the cotton in one respect, or to a piece of calico in the other. In fact, wicks are sometimes made of a kind of wire gauze. You will observe that this vessel is a porous thing; for if I pour a little water on to the top, it will run out at the bottom. You would be puzzled for a good while if I asked you what the state of this vessel is, what is inside it, and why is it there? The vessel is full of water, and yet you see the

---

[5]   The late Duke of Sussex was, we believe, the first to show that a prawn might be washed upon this principle. If the tail, after pulling off the fan part, be placed in a tumbler of water, and the head be allowed to hang over the outside, the water will be sucked up the tail by capillary attraction, and will continue to run out through the head until the water in the glass has sunk so low that the tail ceases to dip into it.

water goes in and runs out as if it were empty. In order to prove this to you, I have only to empty it. The reason is this: the wire, being once wetted, remains wet; the meshes are so small that the fluid is attracted so strongly from the one side to the other, as to remain in the vessel, although it is porous. In like manner, the particles of melted tallow ascend the cotton and get to the top; other particles then follow by their mutual attraction for each other, and as they reach the flame they are gradually burned.

FIG. 56

Here is another application of the same principle. You see this bit of cane. I have seen boys about the streets, who are very anxious to appear like men, take a piece of cane, and light it, and smoke it, as an imitation of a cigar. They are enabled to do so by the permeability of the cane in one direction, and by its capillarity. If I place this piece of cane on a plate containing some camphene (which is very much like paraffine in its general character), exactly in the same manner as the blue fluid rose through the salt will this fluid rise through the piece of cane. There being no pores at the side, the fluid can not go in that direction, but must pass through its length. Already the fluid is at the top of the cane; now I can light it and make it serve as a candle. The fluid has risen by the capillary attraction of the piece of cane, just as it does through the cotton in the candle.

Now the only reason why the candle does not burn all down the side of the wick is that the melted tallow extinguishes the flame. You know that a candle, if turned upside down, so as to allow the fuel hot enough to burn, as it does above, where it is carried in small quantities into the wick, and has all the effect of the heat exercised upon it.

There is another condition which you must learn as regards the candle, without which you would not be able fully to understand the philosophy of it, and that is the vaporous condition of the fuel. In order that you may understand that, let me show you a very pretty but very commonplace experiment. If you blow a candle out cleverly, you will see the vapor rise from it. You have, I know, often smelt the vapor of a blown-out candle, and a very bad smell it is; but if you blow it out cleverly you will be able to see pretty well the vapor into which this solid matter is transformed. I will blow out one of these candles in such a way as not to disturb the air around it by the continuing action of my breath; and now, if I hold a lighted taper two or three inches from the wick, you will observe a train of fire going through the air till it reaches the candle. I am obliged to be quick and ready, because if I allow the vapor time to cool, it becomes condensed into a liquid or solid, or the stream of combustible matter gets disturbed.

Now as to the shape or form of the flame. It concerns us much to know about the condition which the matter of the candle finally assumes at the top of the wick, where you have such beauty and brightness as nothing but combustion or flame can produce. You have the glittering beauty of gold and silver, and the still higher lustre of jewels like the ruby and diamond; but none of these rival the brilliancy and beauty of flame. What diamond can shine like flame? It owes its lustre at night-time to the very flame shining upon it. The flame shines in darkness, but the light which the diamond has is as nothing

Fig. 57

until the flame shines upon it, when it is brilliant again. The candle alone shines by itself and for itself, or for those who have arranged the materials. Now let us look a little at the form of the flame as you see it under the glass shade. It is steady and equal, and its general form is that which is represented in the diagram, varying with atmospheric disturbances, and also varying according to the size of the candle. It is a bright oblong, brighter at the top than toward the bottom, with the wick in the middle, and, besides the wick in the middle, certain darker parts towards the bottom, where the ignition is not so perfect as in the part above. I have a drawing here, sketched many years ago by Hooker, when he made his investigations. It is the drawing of the flame of a lamp, but it will apply to the flame of a candle. The cup of the candle is the vessel or lamp; the melted spermaceti is the oil; and the wick is common to both. Upon that he sets this little flame, and then he represents what is true, a certain quantity of matter rising about it which you do not see, and which, if you have not been here before, or are not familiar with the subject, you will not know of. He has here represented the parts of the surrounding atmosphere that are very essential to the flame, and that are always present with it. There is a current formed, which draws the flame out; for the flame which you see is really drawn out by the current, and drawn upward to a great height, just as Hooker has here shown you by that prolongation of the current in the diagram. You may see this by taking a lighted candle, and putting it in the sun so as to get its shadow thrown on a piece of paper. How remarkable it is that that thing which is light enough to produce shadows of other objects can be made to throw its own shadow on a piece of white paper or card, so that you can actually see streaming round the flame something which is not part of the flame, but is ascending and drawing the flame upward. Now I am going to imitate the sunlight by applying the voltaic battery to the electric lamp. You now see our sun and its great luminosity; and by placing a candle between it and the screen, we get the shadow of the wick; then there is a darkish part, as represented in the diagram, and then a part which is more distinct. Curiously enough, however, what we see in the shadow as the darkest part of the flame is, in reality, the brightest part; and here you see streaming upward the ascending current of hot air, as shown by Hooker, which draws out the flame, supplies it with air, and cools the sides of the cup of melted fuel.

FIG. 58

I can give you here a little farther illustration, for the purpose of showing you how flame goes up or down according to the current. I have here a flame—it is not a candle flame—but you can, no doubt, by this time generalize enough to be able to compare one thing with another. What I am about to do is to change the ascending current that takes the flame upward into a descending current. This I can easily do by the little apparatus you see before me. The flame, as I have said, is not a candle flame, but it is produced by alcohol, so that it shall not smoke too much. I will also color the flame with another substance[6], so that you may trace its course; for, with the spirit alone, you could hardly see well enough to have the opportunity of tracing its direction. By lighting this spirit of wine we have then a flame produced, and you observe that when held in the air it naturally goes upward. You understand now, easily enough, why flames go up under ordinary circumstances: it is because of the draught of air by which the combustion is formed. But now, by blowing the flame down, you see I am enabled to make it go downward into this little chimney, the direction of the current being changed. Before we have concluded this course of lectures we shall show you a lamp in which the flame goes up and the smoke goes down, or the flame goes down and the smoke goes up. You see, then, that we have the power in this way of varying the flame in different directions.

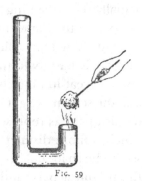

FIG. 59

There are now some other points that I must bring before you. Many of the flames you see here vary very much in their shape by the currents of air blowing around

---

6   The alcohol had chloride of copper dissolved in it: this produces a beautiful green flame.

them in different directions; but we can, if we like, make flames so that they will look like fixtures, and we can photograph them—indeed, we have to photograph them—so that they become fixed to us, if we wish to find out every thing concerning them. That, however, is not the only thing I wish to mention. If I take a flame sufficiently large, it does not keep that homogeneous, that uniform condition of shape, but it breaks out with a power of life which is quite wonderful. I am about to use another kind of fuel, but one which is truly and

FIG. 60

fairly a representative of the wax or tallow of a candle. I have here a large ball of cotton, which will serve as a wick. And, now that I have immersed it in spirit and applied a light to it, in what way does it differ from an ordinary candle? Why, it differs very much in one respect, that we have a vivacity and power about it, a beauty and a life entirely different from the light presented by a candle. You see those fine tongues of flame rising up. You have the same general disposition of the mass of the flame from below upward; but, in addition to that, you have this remarkable breaking out into tongues which you do not perceive in the case of a candle. Now, why is this? I must explain it to you, because, when you understand that perfectly, you will be able to follow me better in what I have to say hereafter. I suppose some here will have made for themselves the experiment I am going to show you. Am I right in supposing that any body here has played at snapdragon? I do not know a more beautiful illustration of the philosophy of flame, as to a certain part of its history, than the game of snapdragon. First, here is the dish; and let me say, that when you play snapdragon properly you ought to have the dish well warmed; you ought also to have warm plums, and warm brandy, which, however, I have not got. When you have put the spirit into the dish, you have the cup and the fuel; and are not the raisins acting like the wicks? I now throw the plums into the dish, and light the spirit, and you see those beautiful tongues of flame that I refer to. You have the air creeping in over the edge of the dish forming these tongues. Why? Because, through the force of the current and the irregularity of the action of the flame, it can not flow in one uniform stream. The air flows in so irregularly that you have what would otherwise be a single image broken up into a variety of forms, and each of these little tongues has an independent existence of its own. Indeed, I might say, you have here a multitude of independent candles. You must not imagine, because you see these tongues all at once, that the flame is of this particular shape. A flame of that shape is never so at any one time. Never is a body of flame, like that which you just saw rising from the ball, of the shape it appears to you. It consists of a multitude of different shapes, succeeding each other so fast that the eye is only able to take cognizance of them all at once. In former times I purposely analyzed a flame of that general character, and the diagram shows you the different parts of which it is composed. They do not occur all at once; it is only because we see these shapes in such rapid succession that they seem to us to exist all at one time.

It is too bad that we have not got farther than my game of snapdragon; but we must not, under any circumstances, keep you beyond your time. It will be a lesson to me in future to hold you more strictly to the philosophy of the thing than to take up your time so much with these illustrations.

LECTURE II

A CANDLE: BRIGHTNESS OF THE FLAME— AIR NECESSARY FOR COMBUSTION— PRODUCTION OF WATER

WE were occupied the last time we met in considering the general character and arrangement as regards the fluid portion of a candle, and the way in which that fluid got into the place of combustion. You see, when we have a candle burning fairly in a regular, steady atmosphere, it will have a shape something like the one shown in the diagram, and will look pretty uniform, although very curious in its character. And now I have to ask your attention to the means by which we are enabled to ascertain what happens in any particular part of the flame; why it happens; what it does in happening; and where, after all, the whole candle goes to; because, as you know very well, a candle being brought before us and burned, disappears, if burned properly, without the least trace of dirt in the candle stick; and this is a very curious circumstance. In order, then, to examine this candle carefully, I have arranged certain apparatus, the use of which you will see as I go on. Here is a candle; I am about to put the end of this glass tube into the middle of the flame—into that part which old Hooker has represented in the diagram as being rather dark, and which you can see at any time if you will look at a candle carefully, without blowing it about. We will examine this dark part first.

Fig. 61

Now I take this bent glass tube, and introduce one end into that part of the flame, and you see at once that something is coming from the flame, out at the other end of the tube; and if I put a flask there, and leave it for a little while, you will see that something from the middle part of the flame is gradually drawn out, and goes through the tube, and into that flask, and there behaves very differently from what it does in the open air. It not only escapes from the end of the tube, but falls down to the bottom of the flask like a heavy substance, as indeed it is. We find that this is the wax of the candle made into a vaporous fluid—not a gas. (You must learn the difference between a gas and a vapor: a gas remains permanent; a vapor is something that will condense.) If you blow out a candle, you perceive a very nasty smell, resulting from the condensation of this vapor.

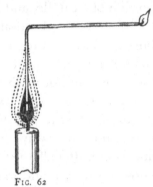

That is very different from what you have outside the flame; and, in order to make that more clear to you, I am about to produce and set fire to a larger portion of this vapor; for what we have in the small way in a candle, to understand thoroughly, we must, as philosophers, produce in a larger way, if needful, that we may examine the different parts.

FIG. 62

And now Mr. Anderson will give me a source of heat, and I am about to show you what that vapor is. Here is some wax in a glass flask, and I am going to make it hot, as the inside of that candle-flame is hot, and the matter about the wick is hot. [The lecturer placed some wax in a glass flask, and heated it over a lamp.] Now I dare say that is hot enough for me. You see that the wax I put in it has become fluid, and there is a little smoke coming from it. We shall very soon have the vapor rising up. I will make it still hotter, and now we get more of it, so that I can actually pour the vapor out of the flask into that basin, and set it on fire there. This, then, is exactly the same kind of vapor as we have in the middle of the candle; and that you may be sure this is the case, let us try whether we have not got here, in this flask, a real combustible vapor out of the middle of the candle. [Taking the flask into which the tube from the candle proceeded, and introducing a lighted taper.] See how it burns. Now this is the vapor from the middle of the candle, produced by its own heat; and that is one of the first things you have to consider with respect to the progress of the wax in the course of its combustion, and as regards the changes it undergoes. I will arrange another tube carefully in the flame, and I should not wonder if we were able, by a little care, to get that vapor to pass through the tube to the other extremity, where we will light it, and obtain absolutely the flame of the candle at a place distant from it. Now, look at that. Is not that a very pretty experiment? Talk about laying on gas—why, we can actually lay on a candle! And you see from this that there are clearly two different kinds of action—one the *production* of the vapor, and the other the *combustion* of it—both of which take place in particular parts of the candle.

I shall get no vapor from that part which is already burnt. If I raise the tube (FIG. 61) to the upper part of the flame, so soon as the vapor has been swept out what comes away will be no longer combustible; it is already burned. How burned? Why, burned thus: In the middle of the flame, where the wick is, there is this combustible vapor; on the outside of the flame is the air which we shall find necessary for the burning of the candle; between the two, intense chemical action takes place, whereby the air and the fuel act upon each other, and at the very same time that we obtain light the vapor inside is destroyed. If you examine where the heat of a candle is, you will find it very curiously arranged. Suppose I take this candle, and hold a piece of paper close upon the flame, where is the heat of that flame? Do you not see that it is *not* in the inside? It is in a ring, exactly in the place where I told you the chemical action was; and even in my irregular mode of making the experiment, if there is not too much disturbance, there will always be a ring. This is a good experiment for you to make at home. Take a strip of paper, have the air in the room quiet, and put the piece of paper right across the middle of the flame—(I must not talk while I make the experiment)—and you will find that it is burnt in two places, and that it is not burnt, or very little so, in the middle; and when you have tried the experiment once or twice, so as to make it nicely, you will be very interested to see where the heat is, and to find that it is where the air and the fuel come together.

This is most important for us as we proceed with our subject. Air is absolutely necessary for combustion; and, what is more, I must have you understand that *fresh* air is necessary, or else we should be imperfect in our reasoning and our experiments. Here is a jar of air; I place it over a candle, and it burns very nicely in it at first, showing that what I have said about it is true; but there will soon be a change. See how the flame is drawing upward, presently fading, and at last going out. And going out, why? Not because it wants air merely, for the jar is as full now as it was before; but it wants pure, fresh air. The jar is full of air, partly changed, partly not changed; but it does not contain sufficient of the fresh air which is necessary for the combustion of a candle. These are all points which we, as young chemists, have to gather up; and if we look a little more closely into this kind of action, we shall find certain steps of reasoning extremely interesting. For instance, here is the oil-lamp I showed you—an excellent lamp for our experiments—the old Argand lamp. I now make it like a candle [obstructing the passage of air into the centre of the flame]; there is the cotton; there is the oil rising up in it, and there is the conical flame. It burns poorly because there is a partial restraint of air. I have allowed no air to get to it save around the outside of the flame, and it does not burn well. I can not admit more air from the outside, because the wick is large; but if, as Argand did so cleverly, I open a passage to the middle of the flame, and so let air come in there, you will see how much more beautifully it burns. If I shut the air off, look how it smokes; and why? We have now some very interesting points to study: we have the case of the combustion of a candle; we have the case of a candle being put out by the want of air; and we have now the case of imperfect combustion, and this is to us so interesting that I want you to understand it as thoroughly as you do the case of a candle burning in its best possible manner. I will now make a great flame, because we need the largest possible illustrations. Here is a larger wick [burning turpentine on a ball of cotton]. All these things are the same as candles, after all. If we have larger wicks, we must have a larger supply of air, or we shall have less perfect combustion. Look, now, at this black substance going up into the atmosphere; there is a regular stream of it. I have provided means to carry off the imperfectly-burned part, lest it should annoy you. Look at the soots that fly off from the flame; see what an imperfect combustion it is, because it can not get enough air. What, then, is happening? Why, certain things which are necessary to the combustion of a candle are absent, and very bad results are accordingly produced; but we see what happens to a candle when it is burnt in a pure and proper state of air. At the time when I showed you this charring by the ring of flame on the one side of the paper, I might have also shown you, by turning to the other side, that the burning of a candle produces the same kind of soot—charcoal, or carbon.

But, before I show that, let me explain to you, as it is quite necessary for our purpose, that, though I take a candle, and give you, as the general result, its combustion in the form of a flame, we must see whether combustion is always in this condition, or whether there are other conditions of flame; and we shall soon discover that there are, and that they are most important to us. I think, perhaps, the best illustration of such a point to us, as juveniles, is to show the result of strong contrast. Here is a little gunpowder. You know that gunpowder burns with flame; we may fairly call it flame. It contains carbon and other materials, which altogether cause it to burn with a flame. And here is some pulverized iron, or iron filings. Now I purpose burning these two things together. I have a little mortar in which I will mix them. (Before I go into these experiments, let me hope that none of you, by trying to repeat them for fun's sake, will do any harm. These things may all be very properly used if you take care, but without that much mischief will be done.) Well, then, here is a little gunpowder, which I put at the bottom of that little wooden vessel, and mix the iron filings up with it, my object being to make the gunpowder set fire to the filings and burn them in the air, and thereby show the difference between substances burning with flame and not with flame. Here is the mixture; and when I set fire to it you must watch the combustion, and you will see that it is of two kinds. You will see the gunpowder burning with a flame and the filings thrown up. You will see them burning, too, but without the production of flame. They will each burn separately. [The lecturer then ignited the mixture.] There is the gunpowder, which burns with a flame, and there are the filings: they burn with a different kind of combustion. You see, then, these two great distinctions; and upon these differences depend all the utility and all the beauty of flame which we use for the purpose of giving out light. When we use oil, or gas, or candle for the purpose of illumination, their fitness all depends upon these different kinds of combustion.

There are such curious conditions of flame that it requires some cleverness and nicety of discrimination to distinguish the kinds of combustion one from another. For instance, here is a powder which is very combustible, consisting, as you see, of separate little particles. It is called *lycopodium*,[7] and each of these particles can produce a vapor, and produce its own flame; but, to see them burning, you would imagine it was all one flame. I will now set fire to a quantity, and you will see the effect. We saw a cloud of flame, apparently in one body; but that rushing noise [referring to the sound produced by the burning] was a proof that the combustion was not a continuous or regular one. This is the lightning of the pantomimes, and a very good imitation. [The experiment was twice repeated by blowing lycopodium from a glass tube

---

[7]   Lycopodium is a yellowish powder found in the fruit of the club moss (*Lycopodium clavatum*). It is used in fireworks.

through a spirit flame.] This is not an example of combustion like that of the filings I have been speaking of, to which we must now return.

Suppose I take a candle and examine that part of it which appears brightest to our eyes. Why, there I get these black particles, which already you have seen many times evolved from the flame, and which I am now about to evolve in a different way. I will take this candle and clear away the gutterage, which occurs by reason of the currents of air; and if I now arrange the glass tube so as just to dip into this luminous part, as in our first experiment, only higher, you see the result. In place of having the same white vapor that you had before, you will now have a black vapor. There it goes, as black as ink. It is certainly very different from the white vapor; and when we put a light to it we shall find that it does not burn, but that it puts the light out. Well, these particles, as I said before, are just the smoke of the candle; and this brings to mind that old employment which Dean Swift recommended to servants for their amusement, namely, writing on the ceiling of a room with a candle. But what is that black substance? Why, it is the same carbon which exists in the candle. How comes it out of the candle? It evidently existed in the candle, or else we should not have had it here. And now I want you to follow me in this explanation. You would hardly think that all those substances which fly about London, in the form of soots and blacks, are the very beauty and life of the flame, and which are burned in it as those iron filings were burned here. Here is a piece of wire gauze, which will not let the flame go through it; and I think you will see, almost immediately, that when I bring it low enough to touch that part of the flame which is otherwise so bright, it quells and quenches it at once, and allows a volume of smoke to rise up.

I want you now to follow me in this point—that whenever a substance burns, as the iron filings burnt in the flame of gunpowder, without assuming the vaporous state (whether it becomes liquid or remains solid), it becomes exceedingly luminous. I have here taken three or four examples apart from the candle on purpose to illustrate this point to you, because what I have to say is applicable to all substances, whether they burn or whether they do not burn—that they are exceedingly bright if they retain their solid state, and that it is to this presence of solid particles in the candle-flame that it owes its brilliancy.

Here is a platinum wire, a body which does not change by heat. If I heat it in this flame, see how exceedingly luminous it becomes. I will make the flame dim for the purpose of giving a little light only, and yet you will see that the heat which it can give to that platinum wire, though far less than the heat it has itself, is able to raise the platinum wire to a far higher state of effulgence. This flame has carbon in it; but I will take one that has no carbon in it. There is a material, a kind of fuel—a vapor, or gas, whichever you like to call it—in that vessel, and it has no solid particles in it; so I take that because it is an example of flame itself burning without any solid matter whatever; and if I now put this solid substance in it, you see what an intense heat it has, and how brightly it causes the solid body to glow. This is the pipe through which we convey this particular gas, which we call hydrogen, and which you shall know all about the next time we meet. And here is a substance called oxygen, by means of which this hydrogen can burn; and although we produce, by their mixture, far greater heat[8] than you can obtain from the candle, yet there is very little light. If, however, I take a solid substance, and put that into it, we produce an intense light. If I take a piece of lime, a substance which will not burn, and which will not vaporize by the heat (and because it does not vaporize remains solid, and remains heated), you will soon observe what happens as to its glowing. I have here a most intense heat produced by

---

[8] Bunsen has calculated that the temperature of the oxyhydrogen blowpipe is 8061° Centigrade. Hydrogen burning in air has a temperature of 3259° C., and coal-gas in air 2350° C.

the burning of hydrogen in contact with the oxygen; but there is as yet very little light—not for want of heat, but for want of particles which can retain their solid state; but when I hold this piece of lime in the flame of the hydrogen as it burns in the oxygen, see how it glows! This is the glorious lime light, which rivals the voltaic light, and which is almost equal to sunlight. I have here a piece of carbon or charcoal, which will burn and give us light exactly in the same manner as if it were burnt as part of a candle. The heat that is in the flame of a candle decomposes the vapor of the wax, and sets free the carbon particles; they rise up heated and glowing as this now glows, and then enter into the air. But the particles, when burnt, never pass off from a candle in the form of carbon. They go off into the air as a perfectly invisible substance, about which we shall know hereafter.

Is it not beautiful to think that such a process is going on, and that such a dirty thing as charcoal can become so incandescent? You see it comes to this—that all bright flames contain these solid particles; all things that burn and produce solid particles, either during the time they are burning, as in the candle or immediately after being burnt, as in the case of the gunpowder and iron filings—all these things give us this glorious and beautiful light.

I will give you a few illustrations. Here is a piece of phosphorus, which burns with a bright flame. Very well; we

Fig. 63

may now conclude that phosphorus will produce, either at the moment that it is burning or afterwards, these solid particles. Here is the phosphorus lighted, and I cover it over with this glass for the purpose of keeping in what is produced. What is all that smoke? That smoke consists of those very particles which are produced by the combustion of the phosphorus. Here again are two substances. This is chlorate of potassa, and this other sulphuret of antimony. I shall mix these together a little, and then they may be burnt in many ways. I shall touch them with a drop of sulphuric acid, for the purpose of giving you an illustration of chemical action, and they will instantly burn.[9] [The lecturer then ignited the mixture by means of sulphuric acid.] Now, from the appearance of things, you can judge for yourselves whether they produce solid matter in burning. I have given you the train of reasoning which will enable you to say whether they do or do not; for what is this bright flame but the solid particles passing off?

Mr. Anderson has in the furnace a very hot crucible. I am about to throw into it some zinc filings, and they will burn with a flame like gunpowder. I make this experiment because you can make it well at home. Now I want you to see what will be the result of the combustion of this zinc. Here it is burning—burning beautifully like a candle, I may say. But what is all that smoke, and what are those little clouds of wool which will come to you if you can not come to them, and make themselves sensible to you in the form of the old philosophic wool, as it was called? We shall have left in that crucible, also, a quantity of this woolly matter. But I will take a piece of this same zinc, and make an experiment a little more closely at home, as it were. You will have here the same thing happening. Here is the piece of zinc; there [pointing to a jet of hydrogen] is the furnace, and we will set to work and try and burn the metal. It glows, you see; there is the combustion; and there is the white substance into which it burns. And so, if I take that flame of hydrogen as the representative of a candle, and show you a substance like zinc burning in the flame, you will

---

9   The following is the action of the sulphuric in inflaming the mixture
     of sulphuret of antimony and chlorate of potassa. A portion of the
     latter is decomposed by the sulphuric acid into oxide of chlorine,
     bisulphate of potassa, and perchlorate of potassa. The oxide of
     chlorine inflames the sulphuret of antimony, which is a combustible
     body, and the whole mass instantly bursts into flame.

see that it was merely during the action of combustion that this substance glowed—while it was kept hot; and if I take a flame of hydrogen and put this white substance from the zinc into it, look how beautifully it glows, and just because it is a solid substance.

I will now take such a flame as I had a moment since, and set free from it the particles of carbon. Here is some camphene, which will burn with a smoke; but if I send these particles of smoke through this pipe into the hydrogen flame you will see they will burn and become luminous, because we heat them a second time. There they are. Those are the particles of carbon reignited a second time. They are those particles which you can easily see by holding a piece of paper behind them, and which, while they are in the flame, are ignited by the heat produced, and, when so ignited, produce this brightness. When the particles are not separated you get no brightness. The flame of coal gas owes its brightness to the separation, during combustion, of these particles of carbon, which are equally in that as in a candle. I can very quickly alter that arrangement. Here, for instance, is a bright flame of gas. Supposing I add so much air to the flame as to cause it all to burn before those particles are set free, I shall not have this brightness; and I can do that in this way: If I place over the jet this wire-gauze cap, as you see, and then light the gas over it, it burns with a non-luminous flame, owing to its having plenty of air mixed with it before it burns; and if I raise the gauze, you see it does not burn below. [The "air-burner," which is of such value in the laboratory, owes its advantage to this principle. It consists of a cylindrical metal chimney, covered at the top with a piece of rather coarse iron wire gauze. This is supported over an Argand burner in such a manner that the gas may mix in the chimney with an amount of air sufficient to burn the carbon and hydrogen simultaneously, so that there may be no separation of carbon in the flame with consequent deposition of soot. The flame, being unable to pass through the wire gauze, burns in a steady, nearly invisible manner above.] There is plenty of carbon in the gas; but, because the atmosphere can get to it, and mix with it before it burns, you see how pale and blue the flame is. And if I blow upon a bright gas flame, so as to consume all this carbon before it gets heated to the glowing point, it will also burn blue. [The lecturer illustrated his remarks by blowing on the gas light.] The only reason why I have not the same bright light when I thus blow upon the flame is that the carbon meets with sufficient air to burn it before it gets separated in the flame in a free state. The difference is solely due to the solid particles not being separated before the gas is burnt.

You observe that there are certain products as the result of the combustion of a candle, and that of these products one portion may be considered as charcoal, or soot; that charcoal, when afterward burnt, produces some other product; and it concerns us very much now to ascertain what that other product is. We showed that something was going away; and I want you now to understand how much is going up into the air; and for that purpose we will have combustion on a little larger scale. From that candle ascends heated air, and two or three experiments will show you the ascending current; but, in order to give you a notion of the quantity of matter which ascends in this way, I will make an experiment by which I shall try to imprison some of the products of this combustion. For this purpose I have here what boys call a fire-balloon; I use

Fig. 64

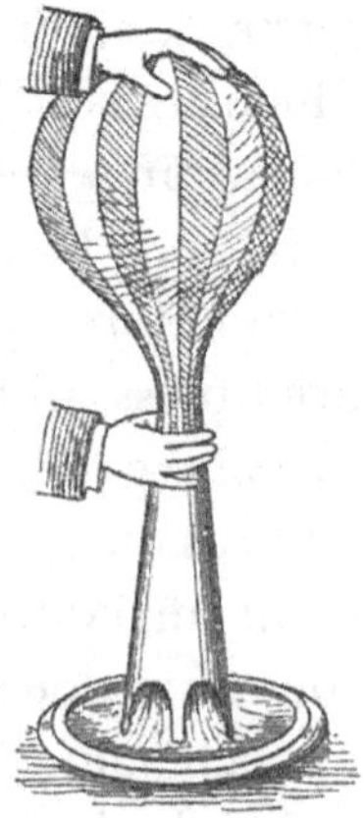

this fire-balloon merely as a sort of measure of the result of the combustion we are considering; and I am about to make a flame in such an easy and simple manner as shall best serve my present purpose. This plate shall be the "cup," we will so say, of the candle; this spirit shall be our fuel; and I am about to place this chimney over it, because it is better for me to do so than to let things proceed at random. Mr. Anderson will now light the fuel, and here at the top we shall get the results of the combustion. What we get at the top of that tube is exactly the same, generally speaking, as you get from the combustion of a candle; but we do not get a luminous flame here, because we use a substance which is feeble in carbon. I am about to put this balloon—not into action, because that is not my object—but to show you the effect which results from the action of those products which arise from the candle, as they arise here from the furnace. [The balloon was held over the chimney (Fig. 64), when it immediately commenced to fill.] You see how it is disposed to ascend; but we must not let it up; because it might come in contact with those upper gaslights, and that would be very inconvenient. [The upper gaslights were turned out at the request of the lecturer, and the balloon was allowed to ascend.] Does not that show you what a large bulk of matter is being evolved? Now there is going through this tube [placing a large glass tube over a candle] all the products of that candle, and you will presently see that the tube will become quite opaque. Suppose I take another candle, and place it under a jar, and then put a light on the other side, just to show you what is going on. You see that the sides of the jar become cloudy, and the light begins to burn feebly. It is the products, you see, which make the light so dim, and this is the same thing which makes the sides of the jar so opaque. If you go home, and take a spoon that has been in the cold air, and hold it over a candle—not so as to soot it—you will find that it becomes dim just as that jar is dim. If you can get a silver dish, or something of that kind, you will make the experiment still better; and now, just to carry your thoughts forward to the time we shall next meet, let me tell you that it is water which causes the dimness, and when we next meet I will show you that we can make it, without difficulty, assume the form of a liquid.

LECTURE III

PRODUCTS: WATER FROM THE COMBUSTION—NATURE OF WATER— A COMPOUND—HYDROGEN

I DARE say you well remember that when we parted we had just mentioned the word "products" from the candle; for when a candle burns we found we were able, by nice adjustment, to get various products from it. There was one substance which was not obtained when the candle was burning properly, which was charcoal or smoke, and there was some other substance that went upward from the flame which did not appear as smoke, but took some other form, and made part of that general current which, ascending from the candle upward, becomes invisible, and escapes. There were also other products to mention. You remember that in that rising current having its origin at the candle we found that one part was condensable against a cold spoon, or against a clean plate, or any other cold thing, and another part was incondensable.

We will first take the condensable part, and examine it, and, strange to say, we find that that part of the product is just water—nothing but water. On the last occasion I spoke of it incidentally, merely saying that water was produced among the condensable products of the candle; but to-day I wish to draw your attention to water, that we may examine it carefully, especially in relation to this subject, and also with respect to its general existence on the surface of the globe.

Now, having previously arranged an experiment for the purpose of condensing water from the products of the candle, my next point will be to show you this water; and perhaps one of the best means that I can adopt for showing its presence to so many at once is to exhibit a very visible action of water, and then to apply that test to what is collected as a drop at the bottom of that vessel. I have here a chemical substance, discovered by

Fig. 65

Sir Humphry Davy, which has a very energetic action upon water, which I shall use as a test of the presence of water. If I take a little piece of it—it is called potassium, as coming from potash—if I take a little piece of it, and throw it into that basin, you see how it shows the presence of water by lighting up and floating about, burning with a violet flame. I am now going to take away the candle which has been burning beneath the vessel containing ice and salt, and you see a drop of water—a condensed product of the candle—hanging from the under surface of the dish (Fig. 65). I will show you that potassium has the same action upon it as upon the water in that basin in the experiment we have just tried. See! it takes fire, and burns in just the same manner. I will take another drop upon this glass slab, and when I put the potassium on to it, you see at once, from its taking fire, that there is water present. Now that water was produced by the candle. In the same manner, if I put this spirit lamp under that jar, you will soon see the latter become damp from the dew which is deposited upon it—that dew being the result of combustion; and I have no doubt you will shortly see, by the drops of water which fall upon the paper below, that there is a good deal of water produced from the combustion of the lamp. I will let it remain, and you can afterward see how much water has been collected. So, if I take a gas lamp, and put any cooling arrangement over it, I shall get water—water being likewise produced from the combustion of gas. Here, in this bottle, is a quantity of water—perfectly pure, distilled water, produced from the combustion of a gas lamp—in no point different from the water that you distill from the river, or ocean, or spring, but exactly the same thing. Water is one individual thing; it never changes. We can add to it by careful adjustment for a little while, or we can take it apart and get other things from it; but water, as water, remains always the same, either in a solid, liquid, or fluid state. Here again [holding another bottle] is some water produced by the combustion of an oil lamp. A pint of oil, when burnt fairly and properly, produces rather more than a pint of water. Here, again, is some water, produced by a rather long experiment, from a wax candle. And so we can go on with almost all combustible substances, and find that if they burn with a flame, as a candle, they produce water. You may make these experiments yourselves: the head of a poker is a very good thing to try with, and if it remains cold long enough over the candle, you may get water condensed in drops on it; or a spoon, or ladle, or any thing else may be used, provided it be clean, and can carry off the heat, and so condense the water.

And now—to go into the history of this wonderful production of water from combustibles, and by combustion—I must first of all tell you that this water may exist in different conditions; and although you may now be acquainted with all its forms, they still require us to give a little attention to them for the present; so that we may perceive how the water, while it goes through its Protean changes, is entirely and absolutely the same thing, whether it is produced from a candle, by combustion, or from the rivers or ocean.

First of all, water, when at the coldest, is ice. Now we philosophers—I hope that I may class you and myself together in this case—speak of water as water, whether it be in its solid, or liquid, or gaseous state—we speak of it chemically as water. Water is a thing compounded of two substances, one of which we have derived from the candle, and the other we shall find elsewhere. Water may occur as ice; and you have had most excellent opportunities lately of seeing this. Ice changes back into water—for we had on our last Sabbath a strong instance of this change by the sad catastrophe which occurred in our own house, as well as in the houses of many of our friends—ice changes back into water when the temperature is raised; water also changes into steam when it is warmed enough. The water which we have here before us is in its densest state [water is in its densest state at a temperature of 39.1° Fahrenheit.]; and, although it changes in weight, in condition, in form, and in many other qualities, it still is water; and whether we alter it into ice by cooling, or whether we change it into steam by heat, it increases in volume—in the one case very strangely and powerfully, and in the other case very largely and wonderfully. For instance, I will now take this tin cylinder, and pour a little water into it, and, seeing how much water I pour in, you may easily estimate for yourselves how high it will rise in the vessel: it will cover the bottom about two inches. I am now about to convert the water into steam for the purpose of showing to you the different volumes which water occupies in its different states of water and steam.

Let us now take the case of water changing into ice: we can effect that by cooling it in a mixture of salt and pounded ice—and I shall do so to show you the expansion of water into a thing of larger bulk when it is so changed. [A mixture of salt and pounded ice reduces the temperature from 32 F. to zero the ice at the same time becoming fluid.] These bottles [holding one] are made of strong cast iron, very strong and very thick—I suppose they are the third of an inch in thickness; they are very carefully filled with water, so as to exclude all air, and then they are screwed down tight. We shall see that when we freeze the water in these iron vessels, they will not be able to hold the ice, and the expansion within them will break them in pieces as these [pointing to some fragments] are broken, which have been bottles of exactly the same kind. I am about to put these two bottles into that mixture of ice and salt for the purpose of showing that when water becomes ice it changes in volume in this extraordinary way.

In the mean time, look at the change which has taken place in the water to which we have applied heat; it is losing its fluid state. You may tell this by two or three circumstances. I have covered the mouth of this glass flask, in which water is boiling, with a watch glass. Do you see what happens? It rattles away like a valve chattering, because the steam rising from the boiling water sends the valve up and down, and forces itself out, and so makes it clatter. You can very easily perceive that the flask is quite full of steam, or else it would not force its way out. You see, also, that the flask contains a substance very much larger than the water, for it fills the whole of the flask over and over again, and there it is blowing away into the air; and yet you can not observe any great diminution in the bulk of the water, which shows you that its change of bulk is very great when it becomes steam.

I have put our iron bottles containing water into this freezing mixture, that you may see what happens. No communication will take place, you observe, between the water in the bottles and the ice in the outer vessel. But there will be a conveyance of heat from the one to the other, and if we are successful—we are making our experiment in very great haste—I expect you will by-and-by, so soon as the cold has taken possession of the bottles and their contents, hear a pop on the occasion of the bursting of the one bottle or the other, and, when we come to examine the bottles, we shall find their contents masses of ice, partly inclosed by the covering of iron which is too small for them, because the ice is larger in bulk than the water. You know very well that ice floats upon water; if a boy falls through a hole into the water, he tries to get on the ice again to float him up. Why does the ice float? Think of that, and philosophize. Because the ice is larger than the quantity of water which can produce it, and therefore the ice weighs the lighter and the water is the heavier.

To return now to the action of heat on water. See what a stream of vapor is issuing from this tin vessel! You observe,

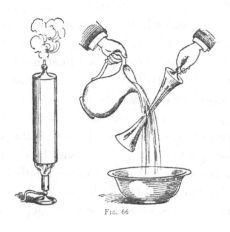

Fig. 66

we must have made it quite full of steam to have it sent out in that great quantity. And now, as we can convert the water into steam by heat, we convert it back into liquid water by the application of cold. And if we take a glass, or any other cold thing, and hold it over this steam, see how soon it gets damp with water: it will condense it until the glass is warm—it condenses the water which is now running down the sides of it. I have here another experiment to show the condensation of water from a vaporous state back into a liquid state, in the same way as the vapor, one of the products of the candle, was condensed against the bottom of the dish and obtained in the form of water; and to show you how truly and thoroughly these changes take place, I will take this tin flask, which is now full of steam, and close the top. We shall see what takes place when we cause this water or steam to return back to the fluid state by pouring some cold water on the outside. [The lecturer poured the cold water over the vessel, when it immediately collapsed (Fig. 66).] You see what has happened. If I had closed the stopper, and still kept the heat applied to it, it would have burst the vessel; yet, when the steam returns to the state of water, the vessel collapses, there being a vacuum produced inside by the condensation of the steam. I show you these experiments for the purpose of pointing out that in all these occurrences there is nothing that changes the water into any other thing; it still remains water; and so the vessel is obliged to give way, and is crushed inward, as in the other case, by the farther application of heat, it would have been blown outward.

And what do you think the bulk of that water is when it assumes the vaporous condition? You see that cube [pointing to a cubic foot]. There, by its side, is a cubic inch (Fig. 67), exactly the same shape as the cubic foot, and that bulk of water [the cubic inch] is sufficient to expand into that bulk [the cubic foot] of steam; and, on the contrary, the application of cold will

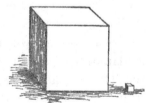

Fig. 67

contract that large quantity of steam into this small quantity of water. [One of the iron bottles burst at that moment.] Ah! There is one of our bottles burst, and here, you see, is a crack down one side an eighth of an inch in width. [The other now exploded, sending the freezing mixture in all directions.] This other bottle is also broken; although the iron was nearly half an inch thick, the ice has burst it asunder. These changes always take place in water; they do not require to be always produced by artificial means; we only use them here because we want to produce a small winter round that little bottle instead of a long and severe one. But if you go to Canada, or to the North, you will find the temperature there out of doors will do the same thing as has been done here by the freezing mixture.

To return to our quiet philosophy. We shall not in future be deceived, therefore, by any changes that are produced in water. Water is the same every where, whether produced from the ocean or from the flame of the candle. Where, then, is this water which we get from a candle? I must anticipate a little, and tell you. It evidently comes, as to part of it, from the candle, but is it within the candle beforehand? No, it is not in the candle; and it is not in the air around about the candle which is necessary for its combustion. It is neither in one nor the other, but it comes from their conjoint action, a part from the candle, a part from the air; and this we have now to trace, so that we may understand thoroughly what is the chemical history of a candle when we have it burning on our table. How shall we get at this? I myself know plenty of ways, but I want you to get at it from the association in your own minds of what I have already told you.

I think you can see a little in this way. We had just now the case of a substance which acted upon the water in the way that Sir Humphry Davy showed us, and which I am now going to recall to your minds by making again an experiment upon that dish. [Potassium, the metallic basis of potash, was discovered by Sir Humphry Davy in 1807, who succeeded in separating it from potash by means of a powerful voltaic battery. Its great affinity for oxygen causes it to decompose water with evolution of hydrogen, which takes fire with the heat produced.] It is a thing which we have to handle very carefully; for you see, if I allow a little splash of water to come upon this mass, it sets fire to part of it; and if there were free access of air, it would quickly set fire to the whole. Now this is a metal—a beautiful and bright metal—which rapidly changes in the air, and, as you know, rapidly changes in water. I will put a piece on the water, and you see it burns beautifully, making a floating lamp, using the water in the place of air. Again, if we take a few iron filings or turnings and put them in water, we find that they likewise undergo an alteration. They do not change so much as this potassium does, but they change somewhat in the same way; they become rusty, and show an action upon the water, though in a different degree of intensity to what this beautiful metal does; but they act upon the water in the same manner generally as this potassium. I want you to put these different facts together in your minds. I have another metal here [zinc], and when we examined it with regard to the solid substance produced by its combustion, we had an opportunity of seeing that it burned; and I suppose, if I take a little strip of this zinc and put it over the candle, you will see something half way, as it were, between the combustion of potassium on the water and the action of iron—you see there is a sort of combustion. It has burned, leaving a white ash or residuum; and here also we find that the metal has a certain amount of action upon water.

By degrees we have learned how to modify the action of these different substances, and to make them tell us what we want to know. And now, first of all, I take iron. It is a common thing in all chemical reactions, where we get any result of this kind, to find that it is increased by the action of heat; and if we want to examine minutely and carefully the action of bodies one upon another, we often have to refer to the action of heat. You are aware, I believe, that iron filings burn beautifully in the air; but I am about to show you an experiment of this

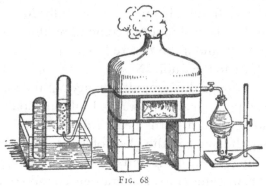

FIG. 68

kind, because it will impress upon you what I am going to say about iron in its action on water. If I take a flame and make it hollow—you know why, because I want to get air to it and into it, and therefore I make it hollow—and then take a few iron filings and drop them into the flame, you see how well they burn. That combustion results from the chemical action which is going on when we ignite those particles. And so we proceed to consider these different effects, and ascertain what iron will do when it meets with water. It will tell us the story so beautifully, so gradually and regularly, that I think it will please you very much.

I have here a furnace with a pipe going through it like an iron gun barrel (Fig. 68), and I have stuffed that barrel full of bright iron turnings, and placed it across the fire to be made red-hot. We can either send air through the barrel to come in contact with the iron, or we can send steam from this little boiler at the end of the barrel. Here is a stop-cock which shuts off the steam from the barrel until we wish to admit it. There is some water in these glass jars, which I have colored blue, so that you may see what happens. Now you know very well that

any steam I might send through that
barrel, if it went through into the
water, would be condensed; for you
have seen that steam can not retain
its gaseous form if it be cooled down;
you saw it here [pointing to the tin
flask] crushing itself into a small
bulk, and causing the flask holding it

FIG. 69

to collapse; so that if I were to send steam through that barrel
it would be condensed, supposing the barrel were cold; it is,
therefore, heated to perform the experiment I am now about
to show you. I am going to send the steam through the barrel
in small quantities, and you shall judge for yourselves, when
you see it issue from the other end, whether it still remains
steam. Steam is condensable into water, and when you lower
the temperature of steam you convert it back into fluid water;
but I have lowered the temperature of the gas which I have
collected in this jar by passing it through water after it has
traversed the iron barrel, and still it does not change back into
water. I will take another test and apply to this gas. (I hold the
jar in an inverted position, or my substance would escape.) If I
now apply a light to the mouth of the jar, it ignites with a slight
noise. That tells you that it is not steam; steam puts out a fire:
it does not burn; but you saw that what I had in that jar burnt.
We may obtain this substance equally from water produced
from the candle flame as from any other source. When it is
obtained by the action of the iron upon the aqueous vapor,
it leaves the iron in a state very similar to that in which these
filings were after they were burnt. It makes the iron heavier
than it was before. So long as the iron remains in the tube
and is heated, and is cooled again without the access of air or
water, it does not change in its weight; but after having had
this current of steam passed over it, it then comes out heavier
than it was before, having taken something out of the steam,
and having allowed something else to pass forth, which we

see here. And now, as we have another jar full, I will show you something most interesting. It is a combustible gas; and I might at once take this jar and set fire to the contents, and show you that it is combustible; but I intend to show you more, if I can. It is also a very light substance. Steam will condense; this body will rise in the air, and not condense. Suppose I take another glass jar, empty of all but air: if I examine it with a taper I shall find that it contains nothing but air. I will now take this jar full of the gas that I am speaking of, and deal with it as though it were a light body; I will hold both upside down, and turn the one up under the other (Fig. 69); and that which did contain the gas procured from the steam, what does it contain now? You will find it now only contains air. But look! Here is the combustible substance [taking the other jar] which I have poured out of the one jar into the other. It still preserves its quality, and condition, and independence, and therefore is the more worthy of our consideration, as belonging to the products of a candle.

Now this substance which we have just prepared by the action of iron on the steam or water, we can also get by means of those other things which you have already seen act so well upon the water. If I take a piece of potassium, and make the necessary arrangements, it will produce this gas; and if, instead, a piece of zinc, I find, when I come to examine it very

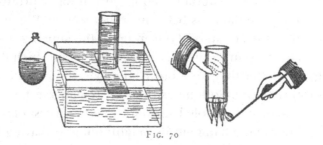

FIG. 70

carefully, that the main reason why this zinc cannot act upon the water continuously as the other metal does, is because the result of the action of the water envelops the zinc in a kind of protecting coat. We have learned in consequence, that if we put into our vessel only the zinc and water, they, by themselves, do not give rise to much action, and we get no result. But suppose I proceed to dissolve off this varnish—this encumbering substance—which I can do by a little acid; the moment I do this I find the zinc acting upon the water exactly as the iron did, but at the common temperature. The acid in no way is altered, except in its combination with the oxide of zinc which is produced. I have now poured the acid into the glass, and the effect is as though I were applying heat to cause this boiling up. There is something coming off from the zinc very abundantly, which is not steam. There is a jar full of it; and you will find that I have exactly the same combustible substance remaining in the vessel, when I hold it upside down, that I produced during the experiment with the iron barrel. This is what we get from water, the same substance which is contained in the candle.

Let us now trace distinctly the connection between these two points. This is hydrogen—a body classed among those things which in chemistry we call elements, because we can get nothing else out of them. A candle is not an elementary body, because we can get carbon out of it; we can get this hydrogen out of it, or at least out of the water which it supplies. And this gas has been so named hydrogen, because it is that element which, in association with another, generates water. Mr. Anderson having now been able to get two or three jars of gas, we shall have a few experiments to make, and I want to show you the best way of making these experiments (Fig. 70). I am not afraid to show you, for I wish you to make experiments, if you will only make them with care and attention, and the assent of those around you. As we advance in chemistry we are obliged to deal with substances which are rather injurious if in their wrong places; the acids, and heat, and combustible things we use, might do harm if carelessly employed. If you want to make hydrogen, you can make it easily from bits of zinc, and sulphuric or muriatic acid. Here is what in former times was called the "philosopher's candle." It is a little phial with a cork and a tube or pipe passing through it. And I am now putting a few little pieces of zinc into it. This little instrument I am going to apply to a useful purpose in our demonstrations, for I want to show you that you can prepare hydrogen, and make some experiments with it as you please, at your own homes. Let me here tell you why I am so careful to fill this phial nearly, and yet not quite full. I do it because the evolved gas, which, as you have seen, is very combustible, is explosive to a considerable extent when mixed with air, and might lead to harm if you were to apply a light to the end of that pipe before all the air had been swept out of the space above the water. I am now about to pour in the sulphuric acid. I have used very little zinc and more sulphuric acid and water, because I want to keep it at work for some time. I therefore take care in this way to modify the proportions of the ingredients so that I may have a regular supply—not too quick and not too slow. Supposing I now take a glass and put it upside down over the end of the tube, because the hydrogen is light I expect that it will remain in that vessel a little while. We will now test the contents of our glass to see if there be hydrogen in it; I think I am safe in saying we have caught some [applying a light]. There it is, you see.

I will now apply a light to the top of the tube. There is the hydrogen burning (Fig. 71). There is our philosophical candle. It is a foolish, feeble sort of a flame, you may say, but it is so hot that scarcely any common flame gives out so much heat. It goes on burning regularly, and I am now about to put that flame to burn under a certain arrangement, in order that we may examine its results and make use of the information which we may thereby acquire. Inasmuch as the candle produces water, and this gas comes out of the water, let us see what this gives us by the same process of combustion that the candle went through

Fig. 71

when it burnt in the atmosphere; and for that purpose I am going to put the lamp under this apparatus (Fig. 72), in order to condense whatever may arise from the combustion within it. In the course of a short time you will see moisture appearing in the cylinder, and you will get the water running down the side, and the water from this hydrogen flame will have absolutely the same effect upon all our tests, being obtained by the same general process as in the former case. This hydrogen is a very beautiful substance. It is so light that it carries things up; it is far lighter than the atmosphere; and I dare say I can show you this by an experiment which if you are very clever, some of you may even have skill enough to repeat. Here is our generator of hydrogen, and here are some soapsuds. I have an India-rubber tube connected with the hydrogen generator, and at the end of the tube is a tobacco pipe. I can thus put the pipe into the suds and blow bubbles by means of the hydrogen. You observe how the bubbles fall downward when I blow them with my warm breath; but notice the difference when I blow them with hydrogen. [The lecturer here blew bubbles with hydrogen, which rose to the roof of the theatre.] It shows you how light this gas must be in order to carry with it not merely the ordinary soap bubble, but the larger portion of a drop hanging to the bottom of it. I can show its lightness in a better way than this; larger bubbles than these may be so lifted up; indeed, in former times balloons used to be filled with this gas. Mr. Anderson will fasten this tube on to our generator, and we shall have a stream of hydrogen here with which we can charge this balloon made of collodion. I need not even be very careful to get all the air out, for I know the power of this gas to carry it up. [Two collodion balloons were inflated and sent up, one being held by a string.] Here is another larger one, made of thin membrane, which we will fill and allow to ascend; you will see they will all remain floating about until the gas escapes.

What, then, are the comparative weights of these substances? I have a table here which will show you the proportion which 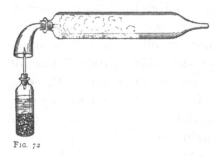 their weights bear to each other. I have taken a pint and a cubic foot as the measures, and have placed opposite to them the respective figures. A pint measure of this hydrogen weighs three-quarters of our smallest weight, a grain, and a cubic foot weighs one-twelfth of an ounce; whereas a pint of water weighs 8,750 grains, and a cubic foot of water weighs almost 1,000 ounces. You see, therefore, what a vast difference there is between the weight of a cubic foot of water and a cubic foot of hydrogen.

FIG. 72

Hydrogen gives rise to no substance that can become solid, either during combustion or afterward as a product of its combustion; but when it burns it produces water only; and if we take a cold glass and put it over the flame, it becomes damp, and you have water produced immediately in appreciable quantity; and nothing is produced by its combustion but the same water which you have seen the flame of the candle produce. It is important to remember that this hydrogen is the only thing in nature which furnishes water as the sole product of combustion.

And now we must endeavor to find some additional proof of the general character and composition of water, and for this purpose I will keep you a little longer, so that at our next meeting we may be better prepared for the subject. We have the power of arranging the zinc which you have seen acting upon the water by the assistance of an acid, in such a manner as to cause all the power to be evolved in the place where we require it. I have behind me a voltaic pile, and I am just about to show you, at the end of this lecture, its character and power, that you may see what we shall have to deal with when next we meet. I hold here the extremities of the wires which transport the power from behind me, and which I shall cause to act on the water.

We have previously seen what a power of combustion is possessed by the potassium, or the zinc, or the iron filings; but none of them show such energy as this. [The lecturer here made contact between the two terminal wires of the battery, when a brilliant flash of light was produced.] This light is, in fact, produced by a forty-zinc power of burning; it is a power that I can carry about in my hands through these wires at pleasure, although if I applied it wrongly to myself, it would destroy me in an instant, for it is a most intense thing, and the power you see here put forth while you count five [bringing the poles in contact and exhibiting the electric light] is equivalent to the power of several thunder-storms, so great is its force. [Professor Faraday has calculated that there is as much electricity required to decompose one grain of water as there is in a very powerful flash of lightning.] And that you may see what intense energy it has, I will take the ends of the wires which convey the power from the battery, and with it I dare say I can burn this iron file. Now this is a chemical power, and one which, when we next meet, I shall apply to water, and show you what results we are able to produce.

J. Clerk Maxwell

1831-1879

Matter and Motion

1877

At age fourteen, J. Clerk Maxwell wrote his first scientific paper, a paper about parabolas. Aside from the tender age of his keen scientific interest, there was nothing special about this paper other than it foreshadowed his many contributions to physics and mathematics. A native of Scotland, he studied at the University of Edinburgh and Cambridge University before going on to teach physics and mathematics at Aberdeen University, King's College London, and Cambridge University where he supervised the establishment of the Cavendish laboratories and edited the Cavendish papers.

Maxwell developed four equations, the Maxwell Equations, which describe the fundamentals of classical electromagnetism. They form the starting point for advanced courses in electromagnetism, and are widely recognized as the greatest contributions to nineteenth-century mathematics. He also showed that the propagation of an electromagnetic field is approximately that of the speed of light. Maxwell was first to propose that light is an electromagnetic phenomenon and that visible light forms only a small part of the whole spectrum of possible electromagnetic radiation. These equations are considered the greatest contributions to nineteenth-century mathematics. His work also set the stage for quantum mechanics and led to modern atomic theory.

Maxwell's research was varied, ranging to astronomy (he discovered that Saturn's rings were made up of particles) and photography (he discovered that color photographs could be created by using only red, green, and blue filters). He also had a profound knowledge of literature and the arts as well as the sciences. Although Maxwell was religious enough to be able to quote the *Psalms* from memory, he saw no reason to confuse faith and science. He once said that no one should try to use science to prove the existence of a higher being; there was no need to do so.

In honor of his outstanding contributions to knowledge, especially his many achievements in physics and mathematics, the largest astronomical telescope in the world, located in Mauna Kea, Hawaii, bears his name.

Sources:

Hutchinson, Ian. 2006. "James Clerk Maxwell and the Christian Proposition." Retrieved February 3, 2007, from http://silas.psfc.mit.edu/maxwell/.

"James Clerk Maxwell." 1997. School of Mathematics and Statistics, University of St. Andrews, Scotland. Retrieved February 3, 2007, from http://www-groups.dcs.st-and.ac.uk/~history/Biographies/Maxwell.html.

"James Clerk Maxwell." 2007. Answers Corporation. Retrieved February 3, 2007, from http://www.answers.com/topic/james-clerk-maxwell.

James Clerk Maxwell Foundation. 2006. Retrieved February 3, 2007, from http://www.clerkmaxwellfoundation.org/.

Maxwell, J. Clerk. 1952. "Biographical Note." *Matter and Motion.* New York: Dover Publications.

Tilanus, Remo. 2007. "James Clerk Maxwell Telescope." Joint Astronomy Centre. Retrieved February 3, 2007, from http://jach.hawaii.edu/JCMT/.

Selection From:

Maxwell, J. Clerk. 1952. *Matter and Motion*. New York: Dover, 1-14.

CHAPTER I

INTRODUCTION

1. NATURE OF PHYSICAL SCIENCE

PHYSICAL SCIENCE is that department of knowledge which relates to the order of nature, or, in other words, to the regular succession of events.

The name of physical science, however, is often applied in a more or less restricted manner to those branches of science in which the phenomena considered are of the simplest and most abstract kind, excluding the consideration of the more complex phenomena, such as those observed in living beings.

The simplest case of all is that in which an event or phenomenon can be described as a change in the arrangement of certain bodies. Thus the motion of the moon may be described by stating the changes in her position relative to the earth in the order in which they follow one another.

In other cases we may know that some change of arrangement has taken place, but we may not be able to ascertain what that change is.

Thus when water freezes we know that the molecules or smallest parts of the substance must be arranged differently in ice and in water. We also know that this arrangement in ice must have a certain kind of symmetry, because the ice is in the form of symmetrical crystals, but we have as yet no precise knowledge of the actual arrangement of the molecules in ice. But whenever we can completely describe the change of arrangement we have a knowledge, perfect so far as it extends, of what has taken place, though we may still have to learn the necessary conditions under which a similar event will always take place.

Hence the first part of physical science relates to the relative position and motion of bodies.

2.   DEFINITION OF A MATERIAL SYSTEM

In all scientific procedure we begin by marking out a certain region or subject as the field of our investigations. To this we must confine our attention, leaving the rest of the universe out of account till we have completed the investigation in which we are engaged. In physical science, therefore, the first step is to define clearly the material system which we make the subject of our statements. This system may be of any degree of complexity. It may be a single material particle, a body of finite size, or any number of such bodies, and it may even be extended so as to include the whole material universe.

3.   DEFINITION OF INTERNAL AND EXTERNAL

All relations or actions between one part of this system and another are called Internal relations or actions.

Those between the whole or any part of the system and bodies not included in the system are called External relations or actions. These we study only so far as they affect the system itself, leaving their effect on external bodies out of consideration. Relations and actions between bodies not included in the system are to be left out of consideration. We cannot investigate them except by making our system include these other bodies.

4.   DEFINITION OF CONFIGURATION

When a material system is considered with respect to the relative position of its parts, the assemblage of relative positions is called the Configuration of the system.

A knowledge of the configuration of the system at a given instant implies a knowledge of the position of every point of the system with respect to every other point at that instant.

5. DIAGRAMS

The configuration of material systems may be represented in models, plans, or diagrams. The model or diagram is supposed to resemble the material system only in form, not necessarily in any other respect.

A plan or a map represents on paper in two dimensions what may really be in three dimensions, and can only be completely represented by a model. We shall use the term Diagram to signify any geometrical figure, whether plane or not, by means of which we study the properties of a material system. Thus, when we speak of the configuration of a system, the image which we form in our minds is that of a diagram, which completely represents the configuration, but which has none of the other properties of the material system. Besides diagrams of configuration we may have diagrams of velocity, of stress, etc., which do not represent the form of the system, but by means of which its relative velocities or its internal forces may be studied.

6. A MATERIAL PARTICLE

A body so small that, *for the purposes of our investigation*, the distances between its different parts may be neglected, is called a material particle.

Thus in certain astronomical investigations the stars, and even the sun, may be regarded each as a material particle, because the difference of the actions of different parts of these bodies does not come under our notice. But we cannot treat them as material particles when we investigate their rotation. Even an atom, when we consider it as capable of rotation, must be regarded as consisting of many material particles.

The diagram of a material particle is of course a mathematical point, which has no configuration.

7.   RELATIVE POSITION OF TWO MATERIAL PARTICLES

The diagram of two material particles consists of two points, as, for instance, A and B.

The position of B relative to A is indicated by the direction and length of the straight line $\overline{AB}$ drawn *from* A to B. If you start from A and travel in the direction indicated by the line $\overline{AB}$ and for a distance equal to the length of that line, you will get to B. This direction and distance may be indicated equally well by any other line, such as ab, which is parallel and equal to $\overline{AB}$. The position of A with respect to B is indicated by the direction and length of the line $\overline{BA}$, drawn from B to A, or the line Ca, equal and parallel to $\overline{BA}$.

It is evident that $\overline{BA} = - \overline{AB}$.

In naming a line by the letters at its extremities, the order of the letters is always that in which the line is to be drawn.

8.   VECTORS

The expression $\overline{AB}$, in geometry, is merely the name of a line. Here it indicates the operation by which the line is drawn, that of carrying a tracing point in a certain direction for a certain distance. As indicating an operation, $\overline{AB}$ is called a Vector, and the operation is completely defined by the direction and distance of the transference. The starting point, which is called the Origin of the vector, may be anywhere.

To define a finite straight line we must state its origin as well as its direction and length. All vectors, however, are regarded as equal which are parallel (and drawn towards the same parts) and of the same magnitude.

Any quantity, such, for instance, as a velocity or a force[1], which has a definite direction and a definite magnitude may be treated as a vector, and may be indicated in a diagram by a straight line whose direction is parallel to the vector, and whose length represents, according to a determinate scale, the magnitude of the vector.

9.   SYSTEM OF THREE PARTICLES

Let us next consider a system of three particles.

Its configuration is represented by a diagram of three points, *A, B, C.*

The position of *B* with respect to *A* is indicated by the vector $\overline{AB}$, and that of C with respect to *B* by the vector $\overline{BC}$.

Fig. 1.

It is manifest that from these data, when *A* is known, we can find *B* and then C, so that the configuration of the three points is completely determined.

The position of *C* with respect to *A* is indicated by the vector $\overline{AC}$, and by the last remark the value of $\overline{AC}$ must be deducible from those of $\overline{AB}$ and $\overline{BC}$.

The result of the operation $\overline{AC}$ is to carry the tracing point from *A* to *C*. But the result is the same if the tracing point is carried first from *A* to *B* and then from *B* to *C*, and this is the sum of the operations $\overline{AB} + \overline{BC}$.

---

[1]   A force is more completely specified as a vector localised in its line of action, called by Clifford a rotor; moreover it is only, when the body on which it acts is treated as rigid that the point of application is inessential.

10. ADDITION OF VECTORS

Hence the rule for the addition of vectors may be stated thus:—From any point as origin draw the successive vectors in series, so that each vector begins at the end of the preceding one. The straight line from the origin to the extremity of the series represents the vector which is the sum of the vectors.

The order of addition is indifferent, for if we write $\overline{BC} + \overline{AB}$ the operation indicated may be performed by drawing $\overline{AD}$ parallel and equal to $\overline{BC}$, and then joining $\overline{DC}$, which, by Euclid, I. 33, is parallel and equal to $\overline{AB}$, so that by these two operations we arrive at the point C in whichever order we perform them.

The same is true for any number of vectors, take them in what order we please.

11. SUBTRACTION OF ONE VECTOR FROM ANOTHER

To express the position of C with respect to B in terms of the positions of B and C with respect to A, we observe that we can get from B to C either by passing along the straight line $\overline{BC}$ or by passing from B to A and then from A to C. Hence

$$\overline{BC} = \overline{BA} + \overline{AC}$$
$$= \overline{AC} + \overline{BA} \text{ since the order of addition is indifferent}$$
$$= \overline{AC} - \overline{AB} \text{ since } \overline{AB} \text{ is equal and opposite to } \overline{BA}.$$

Or the vector BC, which expresses the position of C with respect to B, is found by subtracting the vector of B from the vector of C, these vectors being drawn to B and C respectively from any common origin A.

12. ORIGIN OF VECTORS

The positions of any number of particles belonging to a material system may be defined by means of the vectors drawn to each of these particles from some one point. This point is called the origin of the vectors, or, more briefly, the Origin.

This system of vectors determines the configuration of the whole system; for if we wish to know the position of any point B with respect to any other point A, it may be found from the vectors $\overline{OA}$ and $\overline{OB}$ by the equation

$$\overline{AB} = \overline{OB} - \overline{OA}.$$

We may choose any point whatever for the origin, and there is for the present no reason why we should choose one point rather than another. The configuration of the system—that is to say, the position of its parts with respect to each other—remains the same, whatever point be chosen as origin. Many inquiries, however, are simplified by a proper selection of the origin.

13. RELATIVE POSITION OF Two SYSTEMS

If the configurations of two different systems are known, each system having its own origin, and if we then wish to include both systems in a larger system, having, say, the same origin as the first of the two systems, we must ascertain the position of the origin of the second system with respect to that of the first, and we must be able to draw lines in the second system parallel to those in the first.

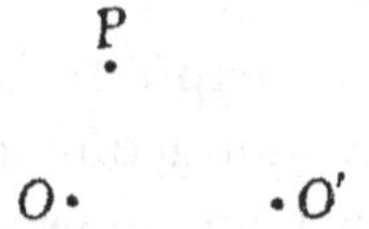

P
.

$O\,\cdot$ $\cdot\,O'$

Fig. 2.

Then by Article 9 the position of a point P of the second system, with respect to the first origin, O, is represented by the sum of the vector $O'P$ of that point with respect to the second origin, $0'$, and the vector $00'$ of the second origin, $0'$, with respect to the first, O.

14. THREE DATA FOR THE COMPARISON OF Two SYSTEMS

We have an instance of this formation of a large system out of two or more smaller systems, when two neighbouring nations, having each surveyed and mapped its own territory, agree to connect their surveys so as to include both countries in one system. For this purpose three things are necessary.

1st. A comparison of the origin selected by the one country with that selected by the other.

2nd. A comparison of the directions of reference used in the two countries.

3rd. A comparison of the standards of length used in the two countries.

1. In civilised countries latitude is always reckoned from the equator, but longitude is reckoned from an arbitrary point, as Greenwich or Paris. Therefore, to make the map of Britain fit that of France, we must ascertain the difference of longitude between the Observatory of Greenwich and that of Paris.

2. When a survey has been made without astronomical instruments, the directions of reference have sometimes been those given by the magnetic compass. This was, I believe, the case in the original surveys of some of the West India islands. The results of this survey, though giving correctly the local configuration of the island, could not be made to fit properly into a general map of the world till the deviation of the magnet from the true north at the time of the survey was ascertained.

3. To compare the survey of France with that of Britain, the metre, which is the French standard of length, must be compared with the yard, which is the British standard of length.

The yard is defined by Act of Parliament 18 and 19 Vict. c. 72, July 30, 1855, which enacts "that the straight line or distance between the centres of the transverse lines in the two gold plugs in the bronze bar deposited in the office of the Exchequer shall be the genuine standard yard at 62° Fahrenheit, and if lost, it shall be replaced by means of its copies."

The metre derives its authority from a law of the French Republic in 1795. It is defined to be the distance between the ends of a certain rod of platinum made by Borda, the rod being at the temperature of melting ice. It has been found by the measurements of Captain Clarke that the metre is equal to 39.37043 British inches.

15.   ON THE IDEA OF SPACE[2]

We have now gone through most of the things to be attended to with respect to the configuration of a material system. There remain, however, a few points relating to the metaphysics of the subject, which have a very important bearing on physics.

We have described the method of combining several configurations into one system which includes them all. In this way we add to the small region which we can explore by stretching our limbs the more distant regions which we can reach by walking or by being carried. To these we add those of which we learn by the reports of others, and those inaccessible regions whose positions we ascertain only by a process of

---

[2]   Following Newton's method of exposition in the *Principia*, a space is assumed and a flux of time is assumed, forming together a framework into which the dynamical explanation of phenomena is set. It is part of the problem of physical astronomy to test this assumption, and to determine this frame with increasing precision. Its philosophical basis can be regarded as a different subject, to which the recent discussions on relativity as regards space and time would be attached. See Appendix I.

calculation, till at last we recognise that every place has a definite position with respect to every other place, whether the one place is accessible from the other or not.

Thus from measurements made on the earth's surface we deduce the position of the centre of the earth relative to known objects, and we calculate the number of cubic miles in the earth's volume quite independently of any hypothesis as to what may exist at the centre of the earth, or in any other place beneath that thin layer of the crust of the earth which alone we can directly explore.

16. ERROR OF DESCARTES

It appears, then, that the distance between one thing and another does not depend on any material thing between them, as Descartes seems to assert when he says (Princip. Phil., II. 18) that if that which is in a hollow vessel were taken out of it without anything entering to fill its place, the sides of the vessel, having nothing between them, would be in contact.

This assertion is grounded on the dogma of Descartes, that the extension in length, breadth, and depth which constitute space is the sole essential property of matter. "The nature of matter," he tells us, "or of body considered generally, does not consist in a thing being hard, or heavy, or coloured, but only in its being extended in length, breadth, and depth" (Princip., II. 4). By thus confounding the properties of matter with those of space, he arrives at the logical conclusion that if the matter within a vessel could be entirely removed, the space within the vessel would no longer exist. In fact he assumes that all space must be always full of matter.

I have referred to this opinion of Descartes in order to show the importance of sound views in elementary dynamics. The primary property of matter was indeed distinctly announced by Descartes in what he calls the "First Law of Nature" (*Princip.*, II.

37): "That every individual thing, so far as in it lies, perseveres in the same state, whether of motion or of rest."[3]

We shall see when we come to Newton's laws of motion that in the words "so far as in it lies," properly understood, is to be found the true primary definition of matter, and the true measure of its quantity. Descartes, however, never attained to a full understanding of his own words (*quantum in se est*), and so fell back on his original confusion of matter with space—space being, according to him, the only form of substance, and all existing things but affections of space. This error[4] runs through every part of Descartes' great work, and it forms one of the ultimate foundations of the system of Spinoza. I shall not attempt to trace it down to more modern times, but I would advise those who study any system of metaphysics to examine carefully that part of it which deals with physical ideas.

We shall find it more conducive to scientific progress to recognise, with Newton, the ideas of time and space as distinct, at least in thought, from that of the material system whose relations these ideas serve to co-ordinate[5].

17. ON THE IDEA OF TIME

The idea of Time in its most primitive form is probably the recognition of an order of sequence in our states of consciousness. If my memory were perfect, I might be able to refer every event within my own experience to its proper place in a chronological series. But it would be difficult, if not impossible, for me to compare the interval between one pair of events and that between another pair—to ascertain, for instance, whether the time during which I can work without feeling tired is greater or less now than when I first began to study. By our intercourse with other persons, and

---

[3] Compare the idea of Least Action: Appendix II.

[4] Some recent forms of relativity have come back to his ideas. Cf. p. 1.40.

[5] See Appendix I.

by our experience of natural processes which go on in a uniform or a rhythmical manner, we come to recognise the possibility of arranging a system of chronology in which all events whatever, whether relating to ourselves or to others, must find their places. Of any two events, say the actual disturbance at the star in Corona Borealis, which caused the luminous effects examined spectroscopically by Mr. Huggins on the 16th May, 1866, and the mental suggestion which first led Professor Adams or M. Leverrier to begin the researches which led to the discovery, by Dr Galle, on the 23rd September, 1846, of the Neptune, the first named must have occurred either before or after the other, or else at the same time.

Absolute, true, and mathematical Time is conceived by Newton as flowing at a constant rate, unaffected by the speed or slowness of the motions of material things. It is also called Duration. Relative, apparent, and common time is duration as estimated by the motion of bodies, as by days, months, and years. These measures of time may be regarded as provisional, for the progress of astronomy has taught us to measure the inequality in the lengths of days, months, and years, and thereby to reduce the apparent time to a more uniform scale, called Mean Solar Time.

18. ABSOLUTE SPACE

Absolute space is conceived as remaining always similar to itself and immovable. The arrangement of the parts of space can no more be altered than the order of the portions of time. To conceive them to move from their places is to conceive a place to move away from itself.

But as there is nothing to distinguish one portion of time from another except the different events which occur in them, so there is nothing to distinguish one part of space from another except its relation to the place of material bodies. We cannot describe the time of an event except by reference to some other event, or the place of a body except by reference to some other body. All our knowledge, both of time and place, is essentially relative[6]. When a man has acquired the habit of putting words together, without troubling himself to form the thoughts which ought to correspond to them, it is easy for him to frame an antithesis between this relative knowledge and a so-called absolute knowledge, and to point out our ignorance of the absolute position of a point as an instance of the limitation of our faculties. Any one, however, who will try to imagine the state of a mind conscious of knowing the absolute position of a point will ever after be content with our relative knowledge.

19. STATEMENT OF THE GENERAL MAXIM OF PHYSICAL SCIENCE

There is a maxim which is often quoted, that "The same causes will always produce the same effects."

To make this maxim intelligible we must define what we mean by the same causes and the same effects, since it is manifest that no event ever happens more than once, so that the causes and effects cannot he the same in *all* respects. What is really meant is that if the causes differ only as regards the absolute time or the absolute place at which the event occurs, so likewise will the effects.

The following statement, which is equivalent to the above maxim, appears to be more definite, more explicitly connected with the ideas of space and time, and more capable of application to particular cases :

"The difference between one event and another does not depend on the mere difference of the times or the places at which they occur, but only on differences in the nature, configuration, or motion of the bodies concerned."

---

6   The position seems to be that our knowledge is relative, but needs definite space and time as a frame for its coherent expression.

It follows from this, that if an event has occurred at a given time and place it is possible for an event exactly similar to occur at any other time and place.

There is another maxim which must not be confounded with that quoted at the beginning of this article, which asserts "That like causes produce like effects."

This is only true when small variations in the initial circumstances produce only small variations in the final state of the system[7]. In a great many physical phenomena this condition is satisfied; but there are other cases in which a small initial variation may produce a very great change in the final state of the system, as when the displacement of the "points" causes a railway train to run into another instead of keeping its proper course.[8]

---

[7] This implies that it is only in so far as stability subsists that principles of natural law can be formulated: it thus perhaps puts a limitation on any postulate of universal physical determinacy such as Laplace was credited with.

[8] We may perhaps say that the observable regularities of nature belong to statistical molecular phenomena which have settled down into permanent stable conditions. In so far as the weather may be clue to an unlimited assemblage of local instabilities, it may not be amenable to a finite scheme of law at all.

EDWIN A. ABBOTT

1838-1926

Flatland

1884

Edwin Abbott was a progressive Anglican clergyman who, in 1884, published a quirky little book titled *Flatland*. A book that is hard to categorize, it is both a brilliant mathematical description of Euclidean space and pure social satire of the Victorian times in which Abbott lived.

Flatland's two-dimensional world is inhabited by flat triangles, squares, higher polygons, and finally perfect circles. They live and move in a planar landscape—with no up or down. When the narrator, A. Square, meets up with a three-dimensional sphere, things really start to happen and not all of it is good. The "Preface," describing Flatland and some of its social structure, is included in this volume.

Source:

Banchoff, Thomas F. 1990. "From Flatland to Hypergraphics: Interacting with Higher Dimensions." *Interdisciplinary Science Reviews*. Retrieved February, 2007, from http://www.geom.uiuc.edu/~banchoff/ISR/ISR.html.

Selection From:

Abbott, Edwin A. 1884. *Flatland.* "Preface." New York: Dover. 1952. 2-17.

PREFACE TO THE SECOND AND REVISED EDITION, 1884. BY THE EDITOR

If my poor Flatland friend retained the vigour of mind which he enjoyed when he began to compose these Memoirs, I should not now need to represent him in this preface, in which he desires, firstly, to return his thanks to his readers and critics in Spaceland, whose appreciation has, with unexpected celerity, required a second edition of his work; secondly, to apologize for certain errors and misprints (for which, however, he is not entirely responsible); and, thirdly, to explain one or two misconceptions. But he is not the Square he once was. Years of imprisonment, and the still heavier burden of general incredulity and mockery, have combined with the natural decay of old age to erase from his mind many of the thoughts and notions, and much also of the terminology, which he acquired during his short stay in Spaceland. He has, therefore, requested me to reply in his behalf to two special objections, one of an intellectual, the other of a moral nature.

The first objection is, that a Flatlander, seeing a Line, sees something that must be thick to the eye as well as long to the eye (otherwise it would not be visible, if it had not some thickness); and consequently he ought (it is argued) to acknowledge that his countrymen are not only long and broad, but also (though doubtless in a very slight degree) thick or high. This objection is plausible, and, to Spacelanders, almost irresistible, so that, I confess, when I first heard it, I knew not what to reply. But my poor old friend's answer appears to me completely to meet it.

"I admit," said he—when I mentioned to him this objection—" I admit the truth of your critic's facts, but I deny his conclusions.

It is true that we have really in Flatland a Third unrecognized Dimension called 'height,' just as it is also true that you have really in Spaceland a Fourth unrecognized Dimension, called by no name at present, but which I will call 'extra-height'. But we can no more take cognizance of our 'height' than you can of your 'extra-height'. Even I—who have been in Spaceland, and have had the privilege of understanding for twenty-four hours the meaning of 'height'—even I cannot now comprehend it, nor realize it by the sense of sight or by any process of reason; I can but apprehend it by faith.

"The reason is obvious. Dimension implies direction, implies measurement, implies the more and the less. Now, all our lines are equally and infinitesimally thick (or high, whichever you like); consequently, there is nothing in them to lead our minds to the conception of that Dimension. No 'delicate micrometer'—as has been suggested by one too hasty Spaceland critic—would in the least avail us; for we should not know what to measure, nor in what direction. When we see a Line, we see something that is long and bright; brightness, as well as length, is necessary to the existence of a Line; if the brightness vanishes, the Line is extinguished. Hence, all my Flatland friends—when I talk to them about the unrecognized Dimension which is somehow visible in a Line—say, 'Ah, you mean brightness': and when I reply, 'No, I mean a real Dimension,' they at once retort 'Then measure it, or tell us in what direction it extends'; and this silences me, for I can do neither. Only yesterday, when the Chief Circle (in other words our High Priest) came to inspect the State Prison and paid me his seventh annual visit, and when for the seventh time he put me the question, 'Was I any better?' I tried to prove to him that he was 'high,' as well as long and broad, although he did not know it. But what was his reply? 'You say I am "high"; measure my "high-ness" and I will believe you.' What could I do? How could I meet his challenge? I was crushed; and he left the room triumphant.

"Does this still seem strange to you? Then put yourself in a similar position. Suppose a person of the Fourth Dimension, condescending to visit you, were to say, 'Whenever you open your eyes, you see a Plane (which is of Two Dimensions) and you infer a Solid (which is of Three); but in reality you also see (though you do not recognize) a Fourth Dimension, which is not colour nor brightness nor anything of the kind, but a true Dimension, although I cannot point out to you its direction, nor can you possibly measure it.' What would you say to such a visitor? Would not you have him locked up? Well, that is my fate: and it is as natural for us Flatlanders to lock up a Square for preaching the Third Dimension, as it is for you Spacelanders to lock up a Cube for preaching the Fourth. Alas, how strong a family likeness runs through blind and persecuting humanity in all Dimensions! Points, Lines, Squares, Cubes, Extra-Cubes—we are all liable to the same errors, all alike the Slaves of our respective Dimensional prejudices, as one of your Spaceland poets has said—

'One touch of Nature makes all worlds akin'."[1]

On this point the defence of the Square seems to me to be impregnable. I wish I could say that his answer to the second (or moral) objection was equally clear and cogent. It has been objected that he is a woman-hater; and as this objection has been vehemently urged by those whom Nature's decree has constituted the somewhat larger half of the Spaceland race, I should like to remove it, so far as I can honestly do so. But the Square is so unaccustomed to the use of the moral terminology of Spaceland that I should be doing him an injustice if I were literally to transcribe his defence against this charge. Acting, therefore, as his interpreter and

---

[1]   The Author desires me to add, that the misconception of some of his critics on this matter has induced him to insert in his dialogue with the Sphere, certain remarks which have a bearing on the point in question, and which he had previously omitted as being tedious and unnecessary.

summarizer, I gather that in the course of an imprisonment of seven years he has himself modified his own personal views, both as regards Women and as regards the Isosceles or Lower Classes. Personally, he now inclines to the opinion of the Sphere that the Straight Lines are in many important respects superior to the Circles. But, writing as a Historian, he has identified himself (perhaps too closely) with the views generally adopted by Flatland, and (as he has been informed) even by Spaceland, Historians; in whose pages (until very recent times) the destinies of Women and of the masses of mankind have seldom been deemed worthy of mention and never of careful consideration.

In a still more obscure passage he now desires to disavow the Circular or aristocratic tendencies with which some critics have naturally credited him. While doing justice to the intellectual power with which a few Circles have for many generations maintained their supremacy over immense multitudes of their countrymen, he believes that the facts of Flatland, speaking for themselves without comment on his part, declare that Revolutions cannot always be suppressed by slaughter, and that Nature, in sentencing the Circles to infecundity, has condemned them to ultimate failure—"and herein," he says, "I see a fulfilment of the great Law of all worlds, that while the wisdom of Man thinks it is working one thing, the wisdom of Nature constrains it to work another, and quite a different and far better thing." For the rest, he begs his readers not to suppose that every minute detail in the daily life of Flatland must needs correspond to some other detail in Spaceland; and yet he hopes that, taken as a whole, his work may prove suggestive as well as amusing, to those Spacelanders of moderate and modest minds who—speaking of that which is of the highest importance, but lies beyond experience—decline to say on the one hand, "This can never be," and on the other hand, "It must needs be precisely thus, and we know all about it."

PART I: THIS WORLD

"Be patient, for the world is broad and wide."

§   1.—*Of the Nature of Flatland*

I CALL our world Flatland, not because we call it so, but to make its nature clearer to you, my happy readers, who are privileged to live in Space.

Imagine a vast sheet of paper on which straight Lines, Triangles, Squares, Pentagons, Hexagons, and other figures, instead of remaining fixed in their places, move freely about, on or in the surface, but without the power of rising above or sinking below it, very much like shadows—only hard and with luminous edges—and you will then have a pretty correct notion of my country and countrymen. Alas, a few years ago, I should have said "my universe": but now my mind has been opened to higher views of things.

In such a country, you will perceive at once that it is impossible that there should be anything of what you call a "solid" kind; but I dare say you will suppose that we could at least distinguish by sight the Triangles, Squares, and other figures, moving about as I have described them. On the contrary, we could see nothing of the kind, not at least so as to distinguish one figure from another. Nothing was visible, nor could be visible, to us, except Straight Lines; and the necessity of this I will speedily demonstrate.

Place a penny on the middle of one of your tables in Space; and leaning over it, look down upon it. It will appear a circle.

But now, drawing back to the edge of the table, gradually lower your eye (thus bringing yourself more and more into the condition of the inhabitants of Flatland), and you will find the penny becoming more and more oval to your view; and at last when you have placed your eye exactly on the edge of the table (so that you are, as it were, actually a Flatlander) the penny will then have ceased to appear oval at all, and will have become, so far as you can see, a straight line.

The same thing would happen if you were to treat in the same way a Triangle, or Square, or any other figure cut out of pasteboard. As soon as you look at it with your eye on the edge on the table, you will find that it ceases to appear to you a figure, and that it becomes in appearance a straight line. Take for example an equilateral Triangle—who represents with us a Tradesman of the respectable class. Fig. 1 represents the Tradesman as you would see him while you were bending over him from above; figs. 2 and 3 represent the Tradesman, as you would see him if your eye were close to the level, or all but on the level of the table; and if your eye were quite on the level of the table (and that is how we see him in Flatland) you would see nothing but a straight line.

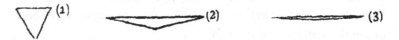

When I was in Spaceland I heard that your sailors have very similar experiences while they traverse your seas and discern some distant island or coast lying on the horizon. The far-off land may have bays, forelands, angles in and out to any number and extent; yet at a distance you see none of these (unless indeed your sun shines bright upon them revealing the projections and retirements by means of light and shade), nothing but a grey unbroken line upon the water.

Well, that is just what we see when one of our triangular or other acquaintances comes toward us in Flatland. As there is neither sun with us, nor any light of such a kind as to make shadows, we have none of the helps to the sight that you have in Spaceland. If our friend comes closer to us we see his line becomes larger; if he leaves us it becomes smaller: but still he looks like a straight line; be he a Triangle, Square, Pentagon, Hexagon, Circle, what you will—a straight Line he looks and nothing else.

You may perhaps ask how under these disadvantageous circumstances we are able to distinguish our friends from one another: but the answer to this very natural question will be more fitly and easily given when I come to describe the inhabitants of Flatland. For the present let me defer this subject, and say a word or two about the climate and houses in our country.

§ 2.—Of the Climate and Houses in Flatland

As with you, so also with us, there arc four points of the compass North, South, East, and West.

There being no sun nor other heavenly bodies, it is impossible for us to determine the North in the usual way; but we have a method of our own. By a Law of Nature with us, there is a constant attraction to the South; and, although in temperate climates this is very slight—so that even a Woman in reasonable health can journey several furlongs northward without much difficulty—yet the hampering effect of the southward attraction is quite sufficient to serve as a compass in most parts of our earth. Moreover, the rain (which falls at stated intervals) coming always from the North, is an additional assistance; and in the towns we have the guidance of the houses, which of course have their side-walls running for the most part North and South, so that the roofs may, keep off the rain from the North. In the country, where there are no houses, the trunks of the trees serve as some sort of guide. Altogether, we have not so much difficulty as might be expected in determining our bearings.

Yet in our more temperate regions, in which the southward attraction is hardly felt, walking sometimes in a perfectly desolate plain where there have been no houses nor trees to guide me, I have been occasionally compelled to remain stationary for hours together, waiting till the rain came before continuing my journey. On the weak and aged, and especially on delicate Females, the force of attraction tells much more heavily than on the robust of the Male Sex, so that it is a point of breeding, if you meet a Lady: in the street, always to give her the North side of the way—by no means an easy thing to do always at short notice when you are in rude health and in a climate where it is difficult to tell your North from your South.

Windows there are none in our houses: for the light comes to us alike in our homes and out of them, by day and by night, equally at all times and in all places, whence we know not. It was in old days, with our learned men, an interesting and oft-investigated question, "What is the origin of light?" and the solution of it has been repeatedly attempted, with no other result than to crowd our lunatic asylums with the would-be solvers. Hence, after fruitless attempts to suppress such investigations indirectly by making them liable to a heavy tax, the Legislature, in comparatively recent times, absolutely prohibited them. I—alas, I alone in Flatland—know now only too well the true solution of this mysterious problem; but my knowledge cannot be made intelligible to a single one of my countrymen; and I am mocked at—I, the sole possessor of the truths of Space and of the theory of the introduction of Light from the world of three Dimensions—as if I were the maddest of the mad! But a truce to these painful digressions: let me return to our houses.

The most common form for the construction of a house is five-sided or pentagonal, as in the annexed figure. The two Northern sides RO, OF, constitute the roof, and for the most part have no doors; on the East is a small door for the Women; on the West a much larger one for the Men; the South side or floor is usually doorless.

Square and triangular houses are not allowed, and for this reason. The angles of a Square (and still more those of an equilateral Triangle,) being much more pointed than those of a Pentagon, and the lines of inanimate objects (such as houses) being dimmer than the lines of Men and Women, it follows that there is no little danger lest the points of a square or triangular house residence might do serious injury to an inconsiderate or perhaps absent-minded traveller suddenly therefore, running against them: and as early as the eleventh century of our era, triangular houses were universally forbidden by Law, the only exceptions being fortifications, powder-magazines, barracks, and other state buildings, which it is not desirable that the general public should. approach without circumspection.

At this period, square houses were still everywhere permitted, though discouraged by a special tax. But, about three centuries afterwards, the Law decided that in all towns containing a population above ten thousand, the angle of a Pentagon was the smallest house-angle that could be allowed consistently with the public safety. The good sense of the community has seconded the efforts of the Legislature; and now, even in the country, the pentagonal construction has superseded every other. It is only now and then in some very remote and backward agricultural district that an antiquarian may still discover a square house.

§ 3.—*Concerning the Inhabitants of Flatland*

The greatest length or breadth of a full grown inhabitant of Flatland may be estimated at about eleven of your inches. Twelve inches may be regarded as a maximum.

Our Women are Straight Lines.

Our Soldiers and Lowest Classes of Workmen are Triangles with two equal sides, each about eleven inches long, and a base or third side so short (often not exceeding half an inch) that they form at their vertices a very sharp and formidable angle. Indeed when their bases are of the most degraded type (not more than the eighth part of an inch in size) they can hardly be distinguished from Straight Lines or Women; so extremely pointed are their vertices. With us, as with you, these Triangles are distinguished from others by being called Isosceles; and by this name I shall refer to them in the following pages.

Our Middle Class consists of Equilateral or Equal-Sided Triangles.

Our Professional Men and Gentlemen are Squares (to which class I myself belong) and Five-Sided Figures or Pentagons.

Next above these come the Nobility, of whom there are several degrees, beginning at Six-Sided Figures, or Hexagons, and from thence rising in the number of their sides till they receive the honourable title of Polygonal, or many-sided. Finally when the number of the sides becomes so numerous, and the sides themselves so small, that the figure cannot be distinguished from a circle, he is included in the Circular or Priestly order; and this is the highest class of all.

It is a Law of Nature with us that a male child shall have one more side than his father, so that each generation shall rise (as a rule) one step in the scale of development and nobility. Thus the son of a Square is a Pentagon; the son of a Pentagon, a Hexagon; and so on.

But this rule applies not always to the Tradesmen, and still less often to the Soldiers, and to the Workmen; who indeed can hardly be said to deserve the name of human Figures, since they have not all their sides equal. With them therefore the Law of Nature does not hold; and the son of an Isosceles (*i.e.* a Triangle with two sides equal) remains Isosceles still. Nevertheless, all hope is not shut out, even from the Isosceles, that his posterity may ultimately rise above his degraded condition. For, after a long series of military successes, or diligent and skilful labours, it is generally found that the more intelligent among the Artisan and Soldier classes manifest a slight increase of their third side or base, and a shrinkage of the two other sides. Intermarriages (arranged by the Priests) between the sons and daughters of these more intellectual members of the lower classes generally result in an offspring approximating still more to the type of the Equal-Sided Triangle.

Rarely—in proportion to the vast numbers of Isosceles births—is a genuine and certifiable Equal-Sided Triangle produced from Isosceles parents.[2] Such a birth requires, as its antecedents, not only a series of carefully arranged intermarriages, but also a long, continued exercise of frugality and self-control on the part of the would-be ancestors of the coming Equilateral, and a patient, systematic, and continuous development of the Isosceles intellect through many generations.

The birth of a True Equilateral Triangle from Isosceles parents is the subject of rejoicing in our country for many furlongs around. After a strict examination conducted by the Sanitary and Social Board, the infant, if certified as Regular, is with solemn ceremonial admitted into the class of Equilaterals. He is then immediately taken from his proud yet sorrowing parents and adopted by some childless Equilateral, who is bound by oath never to permit the child henceforth to enter his former home or so much as to look upon his relations again, for fear lest the freshly developed organism may, by force of unconscious imitation, fall back again into his hereditary level.

The occasional emergence of an Equilateral from the ranks of his serf-born ancestors is welcomed, not only by the poor serfs themselves, as a gleam of light and hope shed upon

---

[2]   "What need of a certificate?" a Spaceland critic may ask: "Is not the procreation of a Square Son a certificate from Nature herself, proving the Equalsidedness of the Father?" I reply that no Lady of any position will marry an uncertified Triangle. Square offspring has sometimes resulted from a slightly Irregular Triangle; but in almost every such case the Irregularity of the first generation is visited on the third; which either fails to attain the Pentagonal rank, or relapses to the Triangular.

the monotonous squalor of their existence, but also by the Aristocracy at large; for all the higher classes are well aware that these rare phenomena, while they do little or nothing to vulgarize their own privileges, serve as a most useful barrier against revolution from below.

Had the acute-angled rabble been all, without exception, absolutely destitute of hope and of ambition, they might have found leaders in some of their many seditious outbreaks, so able as to render their superior numbers and strength too much even for the wisdom of the Circles. But a wise ordinance of Nature has decreed that, in proportion as the working-classes increase in intelligence, knowledge, and all virtue, in that same proportion their acute angle (which makes them physically terrible) shall increase also and approximate to the comparatively harmless angle of the Equilateral Triangle. Thus, in the most brutal and formidable of the soldier class—creatures almost on a level with women in their lack of intelligence—it is found that, as they wax in the mental ability necessary to employ their tremendous penetrating power to advantage, so do they wane in the power of penetration itself.

How admirable is this Law of Compensation! And how perfect a proof of the natural fitness and, I may almost say, the divine origin of the aristocratic constitution of the States in Flatland! By a judicious use of this Law of Nature, the Polygons and Circles are almost always able to stifle sedition in its very cradle, taking advantage of the irrepressible and boundless hopefulness of the human mind. Art also comes to the aid of Law and Order. It is generally found possible—by a little artificial compression or expansion on the part of the State physicians—to make some of the more intelligent leaders of a rebellion perfectly Regular, and to admit them at once into the privileged classes; a much larger number, who are still below the standard, allured by the prospect of being ultimately ennobled, are induced to enter the State Hospitals, where they are kept in honourable confinement for life; one or two alone of the more obstinate, foolish, and hopelessly irregular are led to execution.

Then the wretched rabble of the Isosceles, planless and leaderless, are either transfixed without resistance by the small body of their brethren whom the Chief Circle keeps in pay for emergencies of this kind; or else more often, by means of jealousies and suspicions skilfully fomented among them by the Circular party, they are stirred to mutual warfare, and perish by one another's angles. No less than one hundred and twenty rebellions are recorded in our annals, besides minor outbreaks numbered at two hundred and thirty-five; and they have all ended thus.

§ 4.—*Concerning the Women.*

If our highly pointed Triangles of the Soldier class are formidable, it may be readily inferred that far more formidable are our Women. For if a Soldier is a wedge, a Woman is a needle; being, so to speak, *all* point, at least at the two extremities. Add to this the power of making herself practically invisible at will, and you will perceive that a Female, in Flatland, is a creature by no means to be trifled with.

But here, perhaps, some of my younger Readers may ask *how* a woman in Flatland can make herself invisible. This ought, 1 think, to be apparent without any explanation. However, a few words will make it clear to the most unreflecting.

Place a needle on a table. Then, with your eye on the level of the table, look at it sideways, and you see the whole length of it; but look at it end-ways, and you see nothing but a point, it has become practically invisible. Just so is it with one of our Women, When her side is turned towards us, we see her as a straight line; when the end containing her eye or mouth—for with us these two organs are identical—is the part that meets our eye, then we see nothing but a highly lustrous point; but when the back is presented to our view, then—being only sub-lustrous, and, indeed, almost as dim as an inanimate object—her hinder extremity serves her as a kind of Invisible Cap.

The dangers to which we are exposed from our Women must now be manifest to the meanest capacity in Spaceland. If even the angle of a respectable Triangle in the middle class is not without its dangers; if to run against a Working Man involves a gash; if collision with an officer of the military class necessitates a serious wound; if a mere touch from the vertex of a Private Soldier brings with it danger of death;—what can it be to run against a Woman, except absolute and immediate destruction? And when a Woman is invisible, or visible only as a dim sub-lustrous point, how difficult must it be, even for the most cautious, always to avoid collision!

Many are the enactments made at different times in the different States of Flatland, in order to minimize this peril; and in the Southern and less temperate climates where the force of gravitation is greater, and human beings more liable to casual and involuntary motions, the Laws concerning Women are naturally much more stringent. But a general view of the Code may be obtained from the following summary:—

1. Every house shall have one entrance in the Eastern side, for the use of Females only; by which all females shall enter "in a becoming and respectful manner"[3] and not by the Men's or Western door.
2. No Female shall walk in any public place without continually keeping up her Peace-cry, under penalty of death.
3. Any Female, duly certified to be suffering from St. Vitus's Dance, fits, chronic cold accompanied by violent sneezing, or any disease necessitating involuntary motions, shall be instantly destroyed.

---

[3] When I was in Spaceland I understood that some of your Priestly circles have in the same way a separate entrance for Villagers, Farmers and Teachers of Board Schools (*Spectator*, Sept. 1884, p. 1255) that they may "approach in a becoming and respectful manner."

In some of the States there is an additional Law forbidding Females, under penalty of death, from walking or standing in any public place without moving their backs constantly from right to left so as to indicate their presence to those behind them; others oblige a Woman, when travelling, to be followed by one of her sons, or servants, or by her husband; others confine Women altogether to their houses except during the religious festivals. But it has been found by the wisest of our Circles or Statesmen that the multiplication of restrictions on Females tends not only to the debilitation and diminution of the race, but also to the increase of domestic murders to such an extent that a State loses more than it gains by a too prohibitive Code.

For whenever the temper of the Women is thus exasperated by confinement at home or hampering regulations abroad, they are apt to vent their spleen upon their husbands and children; and in the less temperate climates the whole male population of village has been sometimes destroyed in one or two hours of simultaneous female out break. Hence the Three Laws, mentioned above, suffice for the better regulated States, and may be accepted as a rough exemplification of our Female Code.

After all, our principal safeguard is found, not in Legislature, but in the interests of the Women themselves. For, although they can inflict instantaneous death by a retrograde movement, yet unless they can at once disengage their stinging extremity from the struggling body of their victim, their own frail bodies are liable to be shattered.

The power of Fashion is also on our side. I pointed out that in some less civilized States no female is suffered to stand in any public place without swaying her back from right to left. This practice has been universal among ladies of any pretensions to breeding in all well-governed States, as far back as the memory of Figures can reach. It is considered a disgrace to any State that legislation should have to enforce what ought to be, and is in every respectable female, a natural instinct. The rhythmical and, if I may so say, well-modulated undulation of the back in our ladies of Circular rank is envied and imitated by the wife of a common Equilateral, who can achieve nothing beyond a mere monotonous swing, like the ticking of a pendulum; and the regular tick of the Equilateral is no less admired and copied by the wife of the progressive and aspiring Isosceles, in the females of whose family no "back-motion" of any kind has become as yet a necessity of life. Hence, in every family of position and consideration, "back motion" is as prevalent as time itself; and the husbands and sons in these households enjoy immunity at least from invisible attacks.

Not that it must be for a moment supposed that our Women are destitute of affection. But unfortunately the passion of the moment predominates, in the Frail Sex, over every other consideration. This is, of course, a necessity arising from their unfortunate conformation. For as they have no pretensions to an angle, being inferior in this respect to the very lowest of the Isosceles, they are consequently wholly devoid of brain-power, and have neither reflection, judgment nor forethought, and hardly any memory. Hence, in their fits of fury, they remember no claims and recognize no distinctions. I have actually known a case where a Woman has exterminated her whole household, and half an hour afterwards, when her rage was over and the fragments swept away, has asked what has become of her husband and her children.

Obviously then a Woman is not to be irritated as long as she is in a position where she can turn round. When you have them in their apartments—which are constructed with a view to denying them that power—you can say and do what you like; for they are then wholly impotent for mischief, and will not remember a few minutes hence the incident for which they may be at this moment threatening you with death, nor the promises which you may have found it necessary to make in order to pacify their fury.

On the whole we get on pretty smoothly in our domestic relations, except in the lower strata of the Military Classes.

There the want of tact and discretion on the part of the husbands produces at times indescribable disasters. Relying too much on the offensive weapons of their acute angles instead of the defensive organs of good sense and seasonable simulations, these reckless creatures too often neglect the prescribed construction of the women's apartments, or irritate their wives by ill-advised expressions out of doors, which they refuse immediately to retract. Moreover a blunt and stolid regard for literal truth indisposes them to make those lavish promises by which the more judicious Circle can in a moment pacify his consort. The result is massacre; not, however, without its advantages, as it eliminates the more brutal and troublesome of the Isosceles; and by many of our Circles the destructiveness of the Thinner Sex is regarded as one among many providential arrangements for suppressing redundant population, and nipping Revolution in the bud.

Yet even in our best regulated and most approximately Circular families I cannot say that the ideal of family life is so high as with you in Spaceland. There is peace, in so far as the absence of slaughter may be called by that name, but there is necessarily little harmony of tastes or pursuits; and the cautious wisdom of the Circles has ensured safety at the cost of domestic comfort. In every Circular or Polygonal household it has been a habit from time immemorial—and now has become a kind of instinct among the women of our higher classes—that the mothers and daughters should constantly keep their eyes and mouths towards their husband and his male friends; and for a lady in a family of distinction to turn her back upon her husband would be regarded as a kind of portent, involving loss of *status*. But, as I shall soon shew, this custom, though it has the advantage of safety, is not without its disadvantages.

In the house of the Working Man or respectable Tradesman where the wife is allowed to turn her hack upon her husband while pursuing her household avocations—there are at least intervals of quiet, when the wife is neither seen nor heard, except for the humming sound of the continuous Peace-cry;

but in the homes of the upper classes there is too often no peace. There the voluble mouth and bright penetrating eye are ever directed towards the Master of the household; and light itself is not more persistent than the stream of feminine discourse. The tact and skill which suffice to avert a Woman's sting are unequal to the task of stopping a Woman's mouth; and as the wife has absolutely nothing to say, and absolutely no constraint of wit, sense, or conscience to prevent her from saying it, not a few cynics have been found to aver that they prefer the danger of the death-dealing but inaudible sting to the safe sonorousness of a Woman's other end.

To my readers in Spaceland the condition of our Women may seem truly deplorable, and so indeed it is. A Male of the lowest type of the Isosceles may look forward to some improvement of his angle, and to the ultimate elevation of the whole of his degraded caste; but no Woman can entertain such hopes for her sex. "Once a Woman, always a Woman" is a Decree of Nature; and the very Laws of Evolution seem suspended in her disfavour. Yet at least we can admire the wise Prearrangement which has ordained that, as they have no hopes, so they shall have no memory to recall, and no forethought to anticipate, the miseries and humiliations which are at once a necessity of their existence and the basis of the constitution of Flatland.

René Vallery-Radot

1853-1933

The Life of Pasteur

One of the most important scientists of all time, Louis
Pasteur, produced scientific discoveries that led to the
development of the field of microbiology. His study of disease
advanced germ theory, arguing that microorganisms cause
disease, replacing the belief that diseases were caused by
spontaneous generation. Pasteur developed "pasteurization,"
a process which uses heat to destroy microbes in perishable
foods, such as dairy products. This process made food much
safer and stimulated the huge development of processed
and refrigerated goods, their production, marketing, and
distribution, as well as the home refrigeration industry.
Pasteur also conducted research on rabies, which led to the
development of vaccinations against the dreaded disease.
The excerpt included below describes a pivotal moment in
the validation of his research. It is the dramatic account from
July 1885 when Pasteur used his vaccine to save the life of a
nine-year-old boy, Joseph Meister, who had been bitten by a
rabid dog.

The benefits of Pasteur's work are experienced daily by
humans around the world, from fermentation science and
wine making to improved health and diet to combating deadly

diseases. His academic career and research prepared him well: he earned a BA and BSC degrees from the Royal College in Besançon (France), and from the École normale supérieure in Paris, he earned his an MS in 1845 and his PhD in 1847. In 1848, he became professor of physics at Dijon Lycée, moving then into a more distinguished position in chemistry at the University of Strasbourg in 1849. Pasteur became dean of the new science faculty at Lille University in 1854. In 1888, the Pasteur Institute was established in Paris, headed by Pasteur himself until his death in 1895. Pasteur's biography excerpted below was written by René Vallery-Radot (1853-1933), husband of Pasteur's daughter, Marie-Louise. He served as chairman of the board of the Institute Pasteur 1917-1933.

Selection From:

Vallery-Radot, René. 1924. *The Life of Pasteur.* Trans. R. L. Devonshire. New York: Dover. 1960. 421-432.

CHAPTER XIII

1885-1888

[. . .]

On his return to Paris, Pasteur found himself obliged to hasten the organization of a "service" for the preventive treatment of hydrophobia after a bite. The Mayors of Villers-Farlay, in the Jura, wrote to him that, on October 14, a shepherd had been cruelly bitten by a rabid dog.

Six little shepherd boys were watching over their sheep in a meadow; suddenly they saw a large dog passing along the road, with hanging, foaming jaws.

"A mad dog!" they exclaimed. The dog, seeing the children, left the road and charged them; they ran away shrieking, but the eldest of them, J. B. Jupille, fourteen years of age, bravely turned back in order to protect the flight of his comrades. Armed with his whip, he confronted the infuriated animal, who flew at him and seized his left hand. Jupille, wrestling with the dog, succeeded in kneeling on him, and forcing its jaws open in order to disengage his left hand; in so doing, his right hand was seriously bitten in its turn; finally, having been able to get hold of the animal by the neck, Jupille called to his little brother to pick up his whip, which had fallen during the struggle, and securely fastened the dog's jaws with the lash. He then took his wooden *sabot*, with which he battered the dog's head, after which, in order to be sure that it could do no further harm, he dragged the body down to a little stream in the meadow, and held the head under water for several minutes. Death being now certain, and all danger removed from his comrades, Jupille returned to Villers-Farlay.

Whilst the boy's wounds were being bandaged, the dog's carcase was fetched, and a necropsy took place the next day. The two veterinary surgeons who examined the body had not the slightest hesitation in declaring that the dog was rabid.

The Mayor of Villers-Farlay, who had been to see Pasteur during the summer, wrote to tell him that this lad would die a victim of his own courage unless the new treatment intervened. The answer came immediately: Pasteur declared that, after five years' study, he had succeeded in making dogs refractory to rabies, even six or eight days after being bitten; that he had only once yet applied his method to a human being, but that once with success, in the case of little Meister, and that, if Jupille's family consented, the boy might be sent to him. "I shall keep him near me in a room of my laboratory; he will be watched and need not go to bed; he will merely receive a daily prick, not more painful than a pin-prick."

The family, on hearing this letter, came to an immediate decision; but, between the day when he was bitten and Jupille's arrival in Paris, six whole days had elapsed, whilst in Meister's case there had only been two and a half!

Yet, however great were Pasteur's fears for the life of this tall lad, who seemed quite surprised when congratulated on his courageous conduct, they were not what they had been in the first instance—he felt much greater confidence.

A few days later, on October 26, Pasteur in a statement at the Academy of Sciences described the treatment followed for Meister. Three months and three days had passed, and the child remained perfectly well. Then he spoke of his new attempt. Vulpian rose—

"The Academy will not be surprised," he said, "if, as a member of the Medical and Surgical Section, I ask to be allowed to express the feelings of admiration inspired in me by M. Pasteur's statement. I feel certain that those feelings will be shared by the whole of the medical profession.

"Hydrophobia, that dread disease against which all therapeutic measures had hitherto failed, has at last found a remedy. M. Pasteur, who has been preceded by no one in this path, has been led by a series of investigations unceasingly carried on for several years, to create a method of treatment, by means of which the development of hydrophobia can *infallibly* be prevented in a patient recently bitten by a rabid dog. I say infallibly, because, after what I have seen in M. Pasteur's laboratory, I do not doubt the constant success of this treatment when it is put into full practice a few days only after a rabic bite.

"It is now necessary to see about organizing an installation for the treatment of hydrophobia by M. Pasteur's method. Every person bitten by a rabid dog must be given the opportunity of benefiting by this great discovery, which will seal the fame of our illustrious colleague and bring glory to our whole country."

Pasteur had ended his reading by a touching description of Jupille's action, leaving the Assembly under the impression of that boy of fourteen, sacrificing himself to save his companions. An Academician, Baron Larrey, whose authority was rendered all the greater by his calmness, dignity, and moderation, rose to speak. After acknowledging the importance of Pasteur's discovery, Larrey continued, "The sudden inspiration, agility and courage, with which the ferocious dog was muzzled, and thus made incapable of committing further injury to bystanders, . . . such an act of bravery deserves to be rewarded. I therefore have the honour of begging the Académie des Sciences to recommend to the Académie Française this young shepherd, who, by giving such a generous example of courage and devotion, has well deserved a Montyon prize."

Bouley, then chairman of the Academy, rose to speak in his turn—

"We are entitled to say that the date of the present meeting will remain for ever memorable in the history of medicine, and glorious for French science; for it is that of one of the greatest steps ever accomplished in the medical order of things—a progress realized by the discovery of an efficacious means of preventive treatment for a disease, the incurable nature of which was a legacy handed down by one century to another. From this day, humanity is armed with a means of fighting the fatal disease of hydrophobia and of preventing its onset. It is to M. Pasteur that we owe this, and we could not feel too much admiration or too much gratitude for the efforts on his part which have led to such a magnificent result . . . ."

Five years previously, Bouley, in the annual combined public meeting of the five Academies, had proclaimed his enthusiasm for the discovery of the vaccination of anthrax. But on hearing him again on this October day, in 1885, his colleagues could not but be painfully struck by the change in him; his voice was weak, his race thin and pale. He was dying of an affection of the heart, and quite aware of it, but he was sustained by a wonderful energy, and ready to forget his sufferings in his joy at the thought that the sum of human sorrows would be diminished by Pasteur's victory. He went to the Académic de Médecine the next day to enjoy the echo of the great sitting of the Académie des Sciences. He died on November 29.

The chairman of the Academy of Medicine, M. Jules Bergeron, applauded Pasteur's statement all the more that he too had publicly deplored (in 1862) the impotence of medical science in the presence of this cruel disease.

But while M. Bergeron shared the admiration felt by Vulpian and Dr. Grancher for the experiments which had transformed the rabic virus into its own vaccine, other medical men were divided into several categories: some were full of enthusiasm, others reserved their opinion, many were sceptical, and a few even positively hostile.

As soon as Pasteur's paper was published, people bitten by rabid dogs began to arrive from all sides to the laboratory. The "service" of hydrophobia became the chief business of the day. Every morning was spent by Eugène Viala in preparing the fragments of marrow used for inoculations: in a little room permanently kept at a temperature of 20° to 23° C., stood rows of sterilized flasks, their tubular openings closed by plugs of cotton-wool. Each flask contained a rabic marrow, hanging from the stopper by a thread and gradually drying up by the action of some fragments of caustic potash lying at the bottom of the flask. Viala cut those marrows into small pieces by means of scissors previously put through a flame, and placed them in small sterilized glasses; he then added a few drops of veal broth and pounded the mixture with a glass rod. The vaccinal liquid was now ready; each glass was covered with a paper cover, and bore the date of the medulla used, the earliest of which was fourteen days old. For each patient under treatment from a certain date, there was a whole series of little glasses. Pasteur always attended these operations personally.

In the large hall of the laboratory, Pasteur's collaborators, Messrs. Chamberland and Roux, carried on investigations into contagious diseases under the master's directions; the place was full of flasks, pipets, phials, containing culture broths. Etienne Wasserzug, another curator, hardly more than a boy, fresh from the Ecole Normale, where his bright intelligence and affectionate heart had made him very popular, translated (for he knew the English, German, Italian, Hungarian and Spanish languages, and was awaiting a favourable opportunity of learning Russian) the letters which arrived from all parts of the world; he also entertained foreign scientists. Pasteur had in him a most valuable interpreter. Physicians came from all parts of the world asking to be allowed to study the details of the method. One morning, Dr. Grancher found Pasteur listening to a physician who was gravely and solemnly holding forth his objections to microbian doctrines, and in particular to the treatment of hydrophobia. Pasteur having heard this long monologue, rose and said, "Sir, your language is not very intelligible to me. I am not a physician and do not desire to be one. Never speak to me of your dogma of morbid spontaneity. I am a chemist; I carry out experiments and I try to understand what they teach me. What do you think, doctor?" he added turning to M. Graucher. The latter smilingly answered that the hour for inoculations had struck. They took place at eleven, in Pasteur's study; he, standing by the open door, called out the names of the patients. The date and circumstances of the bites and the veterinary surgeon's certificate were entered in a register, and the patients were divided into series according to the degree of virulence which was to be inoculated on each day of the period of treatment.

Pasteur took a personal interest in each of his patients, helping those who were poor and illiterate to find suitable lodgings in the great capital. Children especially inspired him with a loving solicitude. But his pity was mingled with terror, when, on November 9, a little girl of ten was brought to him who had been severely bitten on the head by a mountain dog, on October 3, thirty-seven days before!! The wound was still suppurating. He said to himself, "This is a hopeless case: hydrophobia is no doubt about to appear immediately; it is much too late for the preventive treatment to have the least chance of success. Should I not, in the scientific interest of the method, refuse to treat this child? If the issue is fatal, all those who have already been treated will be frightened, and many bitten persons, discouraged from coming to the laboratory, may succumb to the disease!" These thoughts rapidly crossed

Pasteur's mind. But he found himself unable to resist his compassion for the father and mother, begging him to try and save their child.

After the treatment was over, Louise Pelletier had returned to school, when fits of breathlessness appeared, soon followed by convulsive spasms; she could swallow nothing. Pasteur hastened to her side when these symptoms began, and new inoculations were attempted. On December 2, there was a respite of a few hours, moments of calm which inspired Pasteur with the vain hope that she might yet be saved. This delusion was a short-lived one. After attending Bouley's funeral, his heart full of sorrow, Pasteur spent the day by little Louise's bedside, in her parents' rooms in the Rue Dauphine. He could not tear himself away; she herself, full of affection for him, gasped out a desire that he should not go away, that he should stay with her! She felt for his hand between two spasms. Pasteur shared the grief of the father and mother. When all hope had to be abandoned: "I did so wish I could have saved your little one!" he said. And as he came down the staircase, he burst into tears.

He was obliged, a few days later, to preside at the reception of Joseph Bertrand at the Académie Française; his sad feelings little in harmony with the occasion. He read in a mournful and troubled voice the speech he had prepared during his peaceful and happy holidays at Arbois. Henry Houssaye, reporting on this ceremony in the *Journal des Débats,* wrote, "M. Pasteur ended his speech amidst a torrent of applause, he received a veritable ovation. He seemed unaccountably moved. How can M. Pasteur, who has received every mark of admiration, every supreme honour, whose name is consecrated by universal renown, still be touched by anything save the discoveries of his powerful genius?" People did not realize that Pasteur's thoughts were far away from himself and from his brilliant discovery. He was thinking of Dumas, his master, of Bouley, his faithful friend and colleague, and of the child he had been unable to snatch from the jaws of death; his mind was not with the living, but with the dead.

A telegram from New York having announced that four children, bitten by rabid dogs, were starting for Paris, many adversaries who had heard of Louise Pelletier's death were saying triumphantly that, if those children's parents had known of her fate, they would have spared them so long and useless a journey.

The four little Americans belonged to workmen's families and were sent to Paris by means of a public subscription opened in the columns of the *New York Herald;* they were accompanied by a doctor and by the mother of the youngest of them, a boy only five years old. After the first inoculation, this little boy, astonished at the insignificant prick, could not help saying, "Is this all we have come such a long journey for?" The children were received with enthusiasm on their return to New York, and were asked "many questions about the great man who had taken such care of them."

A letter dated from that time (January 14, 1886) shows that Pasteur yet found time for kindness, in the midst of his world-famed occupations.

"My dear Jupille, I have received your letters, and I am much pleased with the news you give me of your health. Mme. Pasteur thanks you for remembering her. She, and every one at the laboratory, join with me in wishing that you may keep well and improve as much as possible in reading, writing and arithmetic. Your writing is already much better than it was, but you should take some pains with your spelling. Where do you go to school? Who teaches you? Do you work at home as much as you might? You know that Joseph Meister, who was first to be vaccinated, often writes to me; well, I think he is improving more quickly than you are, though he is only ten years old. So, mind you take pains, do not waste your time with other boys, and listen to the advice of your teachers, and of your father and mother. Remember me to M. Perrot, the Mayor of Villers-Farlay. Perhaps, without him, you would have become ill, and to be ill of hydrophobia means inevitable death; therefore you owe him much gratitude. Good-bye. Keep well."

Pasteur's solicitude did not confine itself to his two first patients, Joseph Meister and the fearless Jupille, but was extended to all those who had come under his care; his kindness was like a living flame. The very little ones who then only saw in him a "kind gentleman" bending over them understood later in life, when recalling the sweet smile lighting up his serious face, that Science, thus understood, unites moral with intellectual grandeur.

Good, like evil, is infectious; Pasteur's science and devotion inspired an act of generosity which was to be followed by many others. He received a visit from one of his colleagues at the Académie Française, Edouard Hervé, who looked upon journalism as a great responsibility and as a school of mutual respect between adversaries. He was bringing to Pasteur, from the Comte de Laubespin, a generous philanthropist, a sum of 40,000 fr. destined to meet the expenses necessitated by the organization of the hydrophobia treatment. Pasteur, when questioned by Hervé, answered that his intention was to found a model establishment in Paris, supported by donations and international subscriptions, without having recourse to the State. But he added that he wanted to wait a little longer until the success of the treatment was undoubted. Statistics came to support it; Bouley, who had been entrusted with an official inquiry on the subject under the Empire, had found that the proportion of deaths after bites from rabid dogs had been 40 per 100, 320 cases having been watched. The proportion often was greater still: whilst Joseph Meister was under Pasteur's care, five persons were bitten by a rabid dog on the Pantin Road, near Paris, and every one of them succumbed to hydrophobia.

Pasteur, instead of referring to Bouley's statistics, preferred to adopt those of M. Leblanc, a veterinary surgeon and a member of the Academy of Medicine, who had for a long time been head of the sanitary department of the *Préfecture de*

Police. These statistics only gave a proportion of deaths of 16 per 100, and had been carefully and accurately kept.

On March 1, he was able to affirm, before the Academy, that the new method had given proofs of its merit, for, out of 350 persons treated, only one death had taken place, that of the little Pelletier. He concluded thus—

"It may be seen, by comparison with the most rigorous statistics, that a very large number of persons have already been saved from death.

"The prophylaxis of hydrophobia after a bite is established.

"It is advisable to create a vaccinal institute against hydrophobia."

The Academy of Sciences appointed a Commission who unanimously adopted the suggestion that an establishment for the preventive treatment of hydrophobia after a bite should be created in Paris, under the name of *Institut Pasteur.* A subscription was about to be opened in France and abroad. The spending of the funds would be directed by a special Committee.

A great wave of enthusiasm and generosity swept from one end of France to another and reached foreign countries. A newspaper of Milan, the *Perseveranza,* which had opened a subscription, collected 6,000 fr. in its first list. The *Journal d'Alsace* headed a propaganda in favour of this work, "sprung from Science and Charity." It reminded its readers that Pasteur had occupied a professor's chair in the former brilliant Faculty of Science of Strasburg, and that his first inoculation was made on an Alsatian boy, Joseph Meister. The newspaper intended to send the subscriptions to Pasteur with these words: "Offerings from Alsace-Lorraine to the Pasteur Institute."

The war of 1870 still darkened the memories of nations. Amongst eager and numerous inventions of instruments of death and destruction, humanity breathed when fresh news came from the laboratory, where a continued struggle was taking place against diseases. The most mysterious, the most cruel of all was going to be reduced to impotence.

Yet the method was about to meet with a few more cases like Louise Pelletier's; accidents would result, either from delay or from exceptionally serious wounds. Happy days were still in store for those who sowed doubt and hatred.

During the early part of March, Pasteur received nineteen Russians, coming from the province of Smolensk. They had been attacked by a rabid wolf and most of them had terrible wounds: one of them, a priest, had been surprised by the infuriated beast as he was going into church, his upper lip and right cheek had been torn off, his face was one gaping wound. Another, the youngest of them, had had the skin of his forehead torn off by the wolf's teeth; other bites were like knife cuts. Five of these unhappy wretches were in such a condition that they had to be carried to the Hôtel Dieu Hospital as soon as they arrived.

The Russian doctor who had accompanied these mujiks related how the wolf had wandered for two days and two nights, tearing to pieces every one he met, and how he had finally been struck down with an axe by one of those he had bitten most severely.

Because of the gravity of the wounds, and in order to make up for the time lost by the Russians before they started, Pasteur decided on making two inoculations every day, one in the morning and one in the evening; the patients at the Hôtel Dieu could be inoculated upon at the hospital.

The fourteen others came every morning in their *touloupes* and fur caps, with their wounds bandaged, and joined without a word the motley groups awaiting treatment at the laboratory—an English family, a Basque peasant, a Hungarian in his national costume, etc., etc.

In the evening, the dumb and resigned band of mujiks came again to the laboratory door. They seemed led by Fate, heedless of the struggle between life and death of which they were the prize. "Pasteur" was the only French word they knew, and their set and melancholy faces brightened in his presence as with a ray of hope and gratitude.

Their condition was the more alarming that a whole fortnight had elapsed between their being bitten and the date of the first inoculations. Statistics were terrifying as to the results of wolf-bites, the average proportion of deaths being 82 per 100. General anxiety and excitement prevailed concerning the hapless Russians, and the news of the death of three of them produced an intense emotion.

Pasteur had unceasingly continued his visits to the Hôtel Dieu. He was overwhelmed with grief. His confidence in his method was in no wise shaken, the general results would not allow it. But questions of statistics were of little account in his eyes when he was the witness of a misfortune; his charity was not of that kind which is exhausted by collective generalities: each individual appealed to his heart. As he passed through the wards at the Hôtel Dieu, each patient in his bed inspired him with deep compassion. And that is why so many who only saw him pass, heard his voice, met his pitiful eyes resting on them, have preserved of him a memory such as the poor had of St. Vincent de Paul.

"The other Russians are keeping well so far," declared Pasteur at the Academy sitting of April 12, 1886. Whilst certain opponents in France continued to discuss the three deaths and apparently saw nought but those failures, the return of the sixteen survivors was greeted with an almost religious emotion. Other Russians had come before them and were saved, and the Tsar, knowing these things, desired his brother, the Grand Duke Vladimir, to bring to Pasteur an imperial gift, the Cross of the Order of St. Anne of Russia, in diamonds. He did more, he gave 100,000 fr. in aid of the proposed Pasteur Institute.

In April, 1886, the English Government, seeing the practical results of the method for the prophylaxis of hydrophobia, appointed a Commission to study and verify the facts. Sir James Paget was the president of it, and the other members were:—Dr. Lauder-Brunton, Mr. Fleming, Sir Joseph Lister, Dr. Quain, Sir Henry Roscoe, Professor Burdon Sanderson, and Mr. Victor Horsley, secretary. The *résumé* of the programme was as follows—

Development of the rabic virus in the medulla oblongata of animals dying of rabies.

Transmission of this virus by subdural or subcutaneous inoculation.

Intensification of this virus by successive passages from rabbit to rabbit.

Possibility either of protecting healthy animals from ulterior bites from rabid animals, or of preventing the onset of rabies in animals already bitten, by means of vaccinal inoculations.

Applications of this method to man and value of its results.

Burdon Sanderson and Horsley came to Paris, and two rabbits, inoculated on by Pasteur, were taken to England; a series of experiments was to be begun on them, and an inquiry was to take place afterwards concerning patients treated both in France and in England. Pasteur, who lost his temper at prejudices and ill-timed levity, approved and solicited inquiry and careful examination.

Long lists of subscribers appeared in the *Journal Officiel*—millionaires, poor workmen, students, women, etc. A great festival was organized at the Trocadéro in favour of the Pasteur Institute; the greatest artistes offered their services. Coquelin recited verses written for the occasion which excited loud applause from the immense audience. Gounod, who had conducted his *Ave Maria*, turned round after the closing bars, and, in an impulse of heartfelt enthusiasm, kissed both his hands to the savant.

In the evening at a banquet, Pasteur thanked his colleagues and the organizers of this incomparable performance. "Was it not," he said, "a touching sight, that of those immortal composers, those great charmers of fortunate humanity coming to the assistance of those who wish to study and to serve suffering humanity? And you too come, great artistes, great actors, like so many generals re-entering the ranks to give greater vigour to a common feeling. I cannot easily describe what I felt. Dare I confess that I was hearing most of you for the first time? I do not think I have spent more than ten evenings of my whole life at a theatre. But I can have no regrets now that you have given me, in a few hours' interval, as in an exquisite synthesis, the feelings that so many others scatter over several months, or rather several years."

A few days later, the subscription from Alsace-Lorraine brought in 43,000 fr. Pasteur received it with grateful emotion, and was pleased and touched to find the name of little Joseph Meister among the list of private subscribers. It was now eleven months since he had been bitten so cruelly by the dog, whose rabic condition had immediately been recognized by the German authorities. Pasteur ever kept a corner of his heart for the boy who had caused him such anxiety.

Marie Curie

1867-1934

Pierre Curie

1923

Marie Sklodowska Curie was the first woman to win the Nobel Prize (1903) and the first person to receive a second Nobel Prize (1911). In 1903, she shared the prize in physics with her husband, Pierre and Henri Becquerel, for the discovery of the phenomenon of radioactivity. Her second prize honored her discovery of the radioactive elements polonium and radium. Her daughter Irene often helped her at the School of Radium at the University of Paris. They worked together during World War I, to create mobile radiography units for the treatment of wounded soldiers. After the war Marie Curie devoted herself to developing the medical uses of radiation. Lifelong she was focused on understanding of radioactivity (a word that she invented) and its potential uses in medicine. In 1935 Irene and her husband, Frederic Joliot, also won the Nobel Prize for their work with radioisotopes.

Marie was born in Warsaw when that city was occupied by the Russians. Her father was an educator who was consistently fired from teaching positions for his fierce patriotism and allegiance to Poland. Although she and her sister, Bronya, wanted to study the sciences at the University of Warsaw, women were not admitted. Instead the sisters attended the "Floating University," an illegal night school in Warsaw, so-called because classes met in changing locations to avoid the authorities. Although the school was not of the caliber of the major universities, she was able to study mathematics, physics, and chemistry. She also worked in private laboratories quietly so that she could learn to conduct experiments, work not sanctioned by the Czarist authorities. She and her sister made a pact that they would help each other through the university by working while the other took classes. Marie worked as governess and tutored children to put her sister through medical school in Paris. Bronya, in turn, helped Marie through her studies at the Sorbonne when she enrolled in 1891. The Sorbonne granted her a master's degree in 1893 and a second degree in math in 1984. In 1903, she received her doctorate from the Sorbonne in physics.

When Marie needed a lab space for her research, a mutual friend suggested that Pierre Curie, who was the laboratory chief at the Municipal School of Industrial Physics and Chemistry in Paris, might have space for her. His laboratory facilities were considered woefully inadequate, but he found a space for Marie—in his lab and in his heart. Their meeting changed their lives; some say it changed the course of science. In a marriage of mutual respect and deep devotion, they committed their research to the study of radioactivity, isolating radioactive elements and discovering radium and polonium by fractionation of pitchblende, leading to much subsequent research in nuclear physics and chemistry. After Pierre was killed on a cold, rainy night when he slipped and fell in front of a horse and buggy, Marie took over his position as head of the Municipal School and later became the first woman professor of science at the Sorbonne. Marie published a biography of her husband, *Pierre Curie,* and at the urging of friends, she included "Autobiographical Notes" at the end of the book. "Chapter One" of those notes is included here.

Marie Curie died in 1934 almost certainly from massive exposure to radiation. All through their research, the Curies were unaware that the handling of this material was seriously jeopardizing their health.

Sources:

"Irene Joliot-Curie."2007. Virtual Museum of Virginia Tech Retrieved August 23, 2007, from http://www.ee.vt. edu/~museum/women/icurie/icurie.html

"Marie Curie and the Science of Radioactivity." 2007. American Institute of Physics. Retrieved February 21, 2007, from http://www.aip.org/history/curie/.

"Marie Curie: The Nobel Prize in Physics 1903." From *Nobel Lectures, Physics 1901-1921*, Elsevier Publishing Company, Amsterdam, 1967. Retrieved February 21, 2007, from http://nobelprize.org/nobel_prizes/physics/laureates/1903/marie-curie-bio.html.

Selection From:

Curie, Marie. 1923. "Autobiographical Notes, Marie Curie: Chapter One." In *Pierre Curie*. Trans. Charlotte and Vernon Kellogg. New York: Macmillan. 155-175.

CHAPTER I

I HAVE been asked by my American friends to write the story of my life. At first, the idea seemed alien to me, but I yielded to persuasion. However, I could not conceive my biography as a complete expression of personal feelings or a detailed description of all incidents I would remember. Many of our feelings change with the years, and, when faded away, may seem altogether strange; incidents lose their momentary interest and may be remembered as if they have occurred to some other person. But there may be in a life some general direction, some continuous thread, due to a few dominant ideas and a few strong feelings, that explain the life and are characteristic of a human personality. Of my life, which has not been easy on the whole, I have described the general course and the essential features, and I trust that my story gives an understanding of the state of mind in which I have lived and worked.

My family is of Polish origin, and my name is Marie Sklodowska. My father and my mother both came from among the small Polish landed proprietors. In my country this class is composed of a large number of families, owners of small and medium-sized estates, frequently interrelated. It has been, until recently, chiefly from this group that Poland has drawn her intellectual recruits.

While my paternal grandfather had divided his time between agriculture and directing a provincial college, my father, more strongly drawn to study, followed the course of the University of Petrograd, and later definitely established himself at Warsaw as Professor of Physics and Mathematics in one of the lyceums of that city. He married a young woman whose mode of life was congenial to his; for, although very young, she had, what was, for that time, a very serious education, and was the director of one of the best Warsaw schools for young girls.

My father and mother worshiped their profession in the highest degree and have left, all over their country, a lasting remembrance with their pupils. I cannot, even to-day, go into Polish society without meeting persons who have tender memories of my parents.

Although my parents adopted a university career, they continued to keep in close touch with their numerous family in the country. It was with their relatives that I frequently spent my vacation, living in all freedom and finding opportunities to know the field life by which I was deeply attracted. To these conditions, so different from the usual villegiature, I believe, I owe my love for the country and nature.

Born at Warsaw, on the 7th of November, 1867, I was the last of five children, but my oldest sister died at the early age of fourteen, and we were left, three sisters and a brother. Cruelly struck by the loss of her daughter and worn away by a grave illness, my mother died at forty-two, leaving her husband in the deepest sorrow with his children. I was then only nine years old, and my eldest brother was hardly thirteen.

This catastrophe was the first great sorrow of my life and threw me into a profound depression. My mother had an exceptional personality. With all her intellectuality she had a big heart and a very high sense of duty. And, though possessing infinite indulgence and good nature, she still held in the family a remarkable moral authority. She had an ardent piety (my parents were both Catholics), but she was never intolerant; differences in religious belief did not trouble her; she was equally kind to any one not sharing her opinions. Her influence over me was extraordinary, for in me the natural love of the little girl for her mother was united with a passionate admiration.

Very much affected by the death of my mother, my father devoted himself entirely to his work and to the care of our education. His professional obligations were heavy and left him little leisure time. For many years we all felt weighing on us the loss of the one who had been the soul of the house.

We all started our studies very young. I was only six years old, and, because I was the youngest and smallest in the class, was frequently brought forward to recite when there were visitors. This was a great trial to me, because of my timidity; I wanted always to run away and hide. My father, an excellent educator, was interested in our work and knew how to direct it, but the conditions of our education were difficult. We began our studies in private schools and finished them in those of the government.

Warsaw was then under Russian domination, and one of the worst aspects of this control was the oppression exerted on the school and the child. The private schools directed by Poles were closely watched by the police and overburdened with the necessity of teaching the Russian language even to children so young that they could scarcely speak their native

Polish. Nevertheless, since the teachers were nearly all of Polish nationality, they endeavored in every possible way to mitigate the difficulties resulting from the national persecution. These schools, however, could not legally give diplomas, which were obtainable only in those of the government.

The latter, entirely Russian, were directly opposed to the Polish national spirit. All instruction was given in Russian, by Russian professors, who, being hostile to the Polish nation, treated their pupils as enemies. Men of moral and intellectual distinction could scarcely agree to teach in schools where an alien attitude was forced upon them. So what the pupils were taught was of questionable value, and the moral atmosphere was altogether unbearable. Constantly held in suspicion and spied upon, the children knew that a single conversation in Polish, or an imprudent word, might seriously harm, not only themselves, but also their families. Amidst these hostilities, they lost all the joy of life, and precocious feelings of distrust and indignation weighed upon their childhood. On the other side, this abnormal situation resulted in exciting the patriotic feeling of Polish youths to the highest degree.

Yet of this period of my early youth, darkened though it was by mourning and the sorrow of oppression, I still keep more than one pleasant remembrance. In our quiet but occupied life, reunions of relatives and friends of our family brought some joy. My father was very interested in literature and well acquainted with Polish and foreign poetry; he even composed poetry himself and was able to translate it from foreign languages into Polish in a very successful way. His little poems on family events were our delight. On Saturday evenings he used to recite or read to us the masterpieces of Polish prose and poetry. These evenings were for us a great pleasure and a source of renewed patriotic feelings.

Since my childhood I have had a strong taste for poetry, and I willingly learned by heart long passages from our great poets, the favorite ones being Mickiewecz, Krasinski and Slowacki. This taste was even more developed when I became acquainted with foreign literatures; my early studies included the knowledge of French, German, and Russian, and I soon became familiar with the fine works written in these languages. Later I felt the need of knowing English and succeeded in acquiring the knowledge of that language and its literature.

My musical studies have been very scarce. My mother was a musician and had a beautiful voice. She wanted us to have musical training. After her death, having no more encouragement from her, I soon abandoned this effort, which I often regretted afterwards.

I learned easily mathematics and physics, as far as these sciences were taken in consideration in the school. I found in this ready help from my father, who loved science and had to teach it himself. He enjoyed any explanation he could give us about Nature and her ways. Unhappily, he had no laboratory and could not perform experiments.

The periods of vacations were particularly comforting, when, escaping the strict watch of the police in the city, we took refuge with relatives or friends in the country. There we found the free life of the old-fashioned family estate; races in the woods and joyous participation in work in the far-stretching, level grain-fields. At other times we passed the border of our Russian-ruled division (Congress Poland) and went southwards into the mountain country of Galicia, where the Austrian political control was less oppressive than that which we suffered. There we could speak Polish in all freedom and sing patriotic songs without going to prison.

My first impression of the mountains was very vivid, because I had been brought up in the plains. So I enjoyed immensely our life in the Carpathian villages, the view of the pikes, the excursions to the valleys and to the high mountain lakes with picturesque names such as: "The Eye of the Sea." However, I never lost my attachment to the open horizon and the gentle views of a plain hill country.

Later I had the opportunity to spend a vacation with my father far more south in Podolia, and to have the first view of the sea at Odessa, and afterwards at the Baltic shore. This was a thrilling experience. But it was in France that I become acquainted with the big waves of the ocean and the ever-changing tide. All my life through, the new sights of Nature made me rejoice like a child.

Thus passed the period of our school life. We all had much facility for intellectual work. My brother, Doctor Sklodowski, having finished his medical studies, became later the chief physician in one of the principal Warsaw hospitals. My sisters and I intended to take up teaching as our parents had done. However, my elder sister, when grown up, changed her mind and decided to study medicine. She took the degree of doctor at the Paris University, married Doctor Dluski, a Polish physician, and together they established an important sanatorium in a wonderfully beautiful Carpathian mountain place of Austrian Poland. My second sister, married in Warsaw, Mrs. Szalay, was for many years a teacher in the schools, where she rendered great service. Later she was appointed in one of the lyceums of free Poland.

I was but fifteen when I finished my high-school studies, always having held first rank in my class. The fatigue of growth and study compelled me to take almost a year's rest in the country. I then returned to my father in Warsaw, hoping to teach in the free schools. But family circumstances obliged me to change my decision. My father, now aged and tired, needed rest; his fortune was very modest. So I resolved to accept a position as governess for several children. Thus, when scarcely seventeen, I left my father's house to begin an independent life.

That going away remains one of the most vivid memories of my youth. My heart was heavy as I climbed into the railway car. It was to carry me for several hours, away from those I loved. And after the railway journey I must drive for five hours longer. What experience was awaiting me? So I questioned as I sat close to the car window looking out across the wide plains.

The father of the family to which I went was an agriculturist. His oldest daughter was about my age, and although working with me, was my companion rather than my pupil. There were two younger children, a boy and a girl. My relations with my pupils were friendly; after our lessons we went together for daily walks. Loving the country, I did not feel lonesome, and although this particular country was not especially picturesque, I was satisfied with it in all seasons. I took the greatest interest in the agricultural development of the estate where the methods were considered as models for the region. I knew the progressive details of the work, the distribution of crops in the fields; I eagerly followed the growth of the plants, and in the stables of the farm I knew the horses.

In winter the vast plains, covered with snow, were not lacking in charm, and we went for long sleigh rides. Sometimes we could hardly see the road. "Look out for the ditch!" I would call to the driver. "You are going straight into it," and "Never fear!" he would answer, as over we went! But these tumbles only added to the gayety of our excursions.

I remember the marvelous snow house we made one winter when the snow was very high in the fields; we could sit in it and look out across the rose-tinted snow plains. We also used to skate on the ice of the river and to watch the weather anxiously, to make sure that the ice was not going to give way, depriving us of our pleasure.

Since my duties with my pupils did not take up all my time, I organized a small class for the children of the village who could not be educated under the Russian government. In this the oldest daughter of the house aided me. We taught the little children and the girls who wished to come how to read and write, and we put in circulation Polish books which were appreciated, too, by the parents. Even this innocent work presented danger, as all initiative of this kind was forbidden by the government and might bring imprisonment or deportation to Siberia.

My evenings I generally devoted to study. I had heard that a few women had succeeded in following certain courses in Petrograd or in foreign countries, and I was determined to prepare myself by preliminary work to follow their example.

I had not yet decided what path I would choose. I was as much interested in literature and sociology as in science. However, during these years of isolated work, trying little by little to find my real preferences, I finally turned towards mathematics and physics, and resolutely undertook a serious preparation for future work. This work I proposed doing in Paris, and I hoped to save enough money to be able to live and work in that city for some time.

My solitary study was beset with difficulties. The scientific education I had received at the lyceum was very incomplete; it was well under the bachelorship program of a French lyceum; I tried to add to it in my own way, with the help of books picked up at random. This method could not be greatly productive, yet it was not without results. I acquired the habit of independent work, and learned a few things which were to be of use later on.

I had to modify my plans for the future when my eldest sister decided to go to Paris to study medicine. We had promised each other mutual aid, but our means did not permit of our leaving together. So I kept my position for three and a half years, and, having finished my work with my pupils, I returned to Warsaw, where a position, similar to the one I had left, was awaiting me.

I kept this new place for only a year and then went back to my father, who had retired some time before and was living alone. Together we passed an excellent year, he occupying himself with some literary work, while I increased our funds by giving private lessons. Meantime I continued my efforts to educate myself. This was no easy task under the Russian government of Warsaw; yet I found more opportunities than in the country. To my great joy, I was able, for the first time in my life, to find access to a laboratory: a small municipal physical laboratory directed by one of my cousins. I found little time to work there, except in the evenings and on Sundays, and was generally left to myself. I tried out various experiments described in treatises on physics and chemistry, and the results were sometimes unexpected. At times I would be encouraged by a little unhoped-for success, at others I would be in the deepest despair because of accidents and failures resulting from my inexperience. But on the whole, though I was taught that the way of progress is neither swift nor easy, this first trial confirmed in me the taste for experimental research in the fields of physics and chemistry.

Other means of instruction came to me through my being one of an enthusiastic group of young men and women of Warsaw, who united in a common desire to study, and whose activities were at the same time social and patriotic. It was one of those groups of Polish youths who believed that the hope of their country lay in a great effort to develop the intellectual and moral strength of the nation, and that such an effort would lead to a better national situation. The nearest purpose was to work at one's own instruction and to provide means of instruction for workmen and peasants. In accordance with this program we agreed among ourselves to give evening courses, each one teaching what he knew best. There is no need to say that this was a secret organization, which made everything extremely difficult. There were in our group very devoted young people who, as I still believe today, could do truly useful work.

I have a bright remembrance of the sympathetic intellectual and social companionship which I enjoyed at that time. Truly the means of action were poor and the results obtained could not be considerable; yet I still believe that the ideas which inspired us then are the only way to real social progress. You cannot hope to build a better world without improving the individuals. To that end each of us must work for his own improvement, and at the same time share a general responsibility for all humanity, our particular duty being to aid those to whom we think we can be most useful.

All the experiences of this period intensified my longing for further study. And, in his affection for me, my father, in spite of limited resources, helped me to hasten the execution of my early project. My sister had just married at Paris, and it was decided that I should go there to live with her. My father and I hoped that, once my studies were finished, we would again live happily together. Fate was to decide otherwise, since my marriage was to hold me in France. My father, who in his own youth had wished to do scientific work, was consoled in our separation by the progressive success of my work. I keep a tender memory of his kindness and disinterestedness. He lived with the family of my married brother, and, like an excellent grandfather, brought up the children. We had the sorrow of losing him in 1902, when he had just passed seventy.

So it was in November, 1891, at the age of twenty-four, that I was able to realize the dream that had been always present in my mind for several years.

When I arrived in Paris I was affectionately welcomed by my sister and brother-in-law, but I stayed with them only for a few months, for they lived in one of the outside quarters of Paris where my brother-in-law was beginning a medical practice, and I needed to get nearer to the schools. I was finally installed, like many other students of my country, in a modest little room for which I gathered some furniture. I kept to this way of living during the four years of my student life.

It would be impossible to tell of all the good these years brought to me. Undistracted by any outside occupation, I was entirely absorbed in the joy of learning and understanding. Yet, all the while, my living conditions were far from easy, my own funds being small and my family not having the means to aid me as they would have liked to do. However, my situation was not exceptional; it was the familiar experience of many of the Polish students whom I knew. The room I lived in was in a garret, very cold in winter, for it was insufficiently heated by a small stove which often lacked coal. During a particularly rigorous winter, it was not unusual for the water to freeze in the basin in the night; to be able to sleep I was obliged to pile all my clothes on the bedcovers. In the same room I prepared my meals with the aid of an alcohol lamp and a few kitchen utensils. These meals were often reduced to bread with a cup of chocolate, eggs or fruit. I had no help in housekeeping and I myself carried the little coal I used up the six flights.

This life, painful from certain points of view, had, for all that, a real charm for me. It gave me a very precious sense of liberty and independence. Unknown in Paris, I was lost in the great city, but the feeling of living there alone, taking care of myself without any aid, did not at all depress me. If sometimes I felt lonesome, my usual state of mind was one of calm and great moral satisfaction.

All my mind was centered on my studies, which, especially at the beginning, were difficult. In fact, I was insufficiently prepared to follow the physical science course at the Sorbonne, for, despite all my efforts, I had not succeeded in acquiring in Poland a preparation as complete as that of the French students following the same course. So I was obliged to supply this deficiency, especially in mathematics. I divided my time between courses, experimental work, and study in the library. In the evening I worked in my room, sometimes very late into the night. All that I saw and learned that was new delighted me. It was like a new world opened to me, the world of science, which I was at last permitted to know in all liberty.

I have pleasant memories of my relations with my student companions. Reserved and shy at the beginning, it was not long before I noticed that the students, nearly all of whom worked seriously, were disposed to be friendly. Our conversations about our studies deepened our interest in the problems we discussed.

Among the Polish students I did not have any companions in my studies. Nevertheless, my relations with their small colony had a certain intimacy. From time to time we would gather in one another's bare rooms, where we could talk over national questions and feel less isolated. We would also go for walks together, or attend public reunions, for we were all interested in politics. By the end of the first year, however, I was forced to give up these relationships, for I found that all my energy had to be concentrated on my studies, in order to achieve them as soon as possible. I was even obliged to devote most of my vacation time to mathematics.

My persistent efforts were not in vain. I was able to make up for the deficiency of my training and to pass examinations at the same time with the other students. I even had the satisfaction of graduating in first rank as "*licenciée es sciences physiques*" in 1893, and in second rank as "*licenciée es sciences mathématiques*" in 1894.

My brother-in-law, recalling later these years of work under the conditions I have just described, jokingly referred to them as "the heroic period of my sister-in-law's life." For myself, I shall always consider one of the best memories of my life that period of solitary years exclusively devoted to the studies, finally within my reach, for which I had waited so long.

It was in 1894 that I first met Pierre Curie. One of my compatriots, a professor at the University of Fribourg, having called upon me, invited me to his home, with a young physicist of Paris, whom he knew and esteemed highly. Upon entering the room I perceived, standing framed by the French window opening on the balcony, a tall young man with auburn hair and large, limpid eyes. I noticed the grave and gentle expression of his face, as well as a certain abandon in his attitude, suggesting the dreamer absorbed in his reflections. He showed me a simple cordiality and seemed to me very sympathetic. After that first interview he expressed the desire to see me again and to continue our conversation of that evening on scientific and social subjects in which he and I were both interested, and on which we seemed to have similar opinions.

Some time later, he came to me in my student room and we became good friends. He described to me his days, filled with work, and his dream of an existence entirely devoted to science. He was not long in asking me to share that existence, but I could not decide at once; I hesitated before a decision that meant abandoning my country and my family.

I went back to Poland for my vacation, without knowing whether or not I was to return to Paris. But circumstances permitted me again to take up my work there in the autumn of that year. I entered one of the physics laboratories at the Sorbonne, to begin experimental research in preparation for my doctor's thesis.

Again I saw Pierre Curie. Our work drew us closer and closer, until we were both convinced that neither of us could find a better life companion. So our marriage was decided upon and took place a little later, in July, 1895.

Pierre Curie had just received his doctor's degree and had been made professor in the School of Physics and Chemistry of the City of Paris. He was thirty-six years old, and already a physicist known and appreciated in France and abroad. Solely preoccupied with scientific investigation, he had paid little attention to his career, and his material resources were very modest. He lived at Sceaux, in the suburbs of Paris, with his old parents, whom he loved tenderly, and whom he described as "exquisite" the first time he spoke to me about them. In fact, they were so: the father was an elderly physician of high intellect and strong character, and the mother the most excellent of women, entirely devoted to her husband and her sons. Pierre's elder brother, who was then professor at the University of Montpellier, was always his best friend. So I had the privilege of entering into a family worthy of affection and esteem, and where I found the warmest welcome.

We were married in the simplest way. I wore no unusual dress on my marriage day, and only a few friends were present at the ceremony, but I had the joy of having my father and my second sister come from Poland.

We did not care for more than a quiet place in which to live and to work, and were happy to find a little apartment of three rooms with a beautiful view of a garden. A few pieces of furniture came to us from our parents. With a money gift from a relative we acquired two bicycles to take us out into the country.

G. H. HARDY

1877-1947

A Mathematician's Apology

1940

G.H. (Godfrey Harold) Hardy showed a proclivity for numbers from his earliest years. Both of his parents were educators with talents for mathematics, so they encouraged his intellectual activity. Graduating from Cambridge University, he took a position at Trinity College as a Fellow (1900) and then a lecturer in mathematics from 1906 to 1919. He was also distinguished as the Cayley lecturer at Cambridge from 1914 to 1919. In 1919, Hardy became a Fellow of New College and Savilian Professor of Geometry at Oxford. He returned to Cambridge in 1931 as Sadleiran Professor of Pure Mathematics and Fellow of Trinity, where he remained until his death.

Most of Hardy's major contributions in mathematics issue from his famous collaboration with John E. Littlewood and Ramanujan, the self-educated mathematical genius from India. Hardy saw his mathematics as beautiful and pure, and he hoped that no one would ever apply them. However, once, in 1908 his mathematics were applied: Hardy and Wilhelm Weinberg, a German physician, framed the Hardy-Weinberg law of genetics, which has to do with the proportions of dominant and recessive traits propagated in a large mixed population. This was the only application of his mathematics. The essay excerpted here from *A Mathematician's Apology*, was written 1940, toward the end of Hardy's career. For those interested in how mathematics is beautiful and important for its own sake, this work is essential reading (selected chapters are included in this volume).

Hardy liked to play tennis and other sports, and he was an avid cricket fan throughout his life. John Maynard Keynes once told him that if he (Hardy) had read the stock market reports for half an hour each day with the same concentration with which he read the cricket scores, he surely would have been a rich man.

Source:

"Godfrey Harold Hardy." 2002. School of Mathematics and Statistics, University of St. Andrews, Scotland. Retrieved February 22, 2007, from http://www-history.mcs.st-andrews.ac.uk/Biographies/Hardy.html.

Selection From:

Hardy, Godfrey H. 1940. *A Mathematician's Apology*. Cambridge: Cambridge University Press. 1967. 63-70, 84-88.

2

I PROPOSE to put forward an apology for mathematics; and I may be told that it needs none, since there are now few studies more generally recognized, for good reasons or bad, as profitable and praiseworthy. This may be true; indeed it is probable, since the sensational triumphs of Einstein, that stellar astronomy and atomic physics are the only sciences which stand higher in popular estimation. A mathematician need not now consider himself on the defensive. He does not have to meet the sort of opposition described by Bradley in the admirable defence of metaphysics which forms the introduction to *Appearance and Reality*.

A metaphysician, says Bradley, will be told that 'metaphysical knowledge is wholly impossible', or that 'even if possible to a certain degree, it is practically no knowledge worth the name'. 'The same problems,' he will hear, 'the same disputes, the same sheer failure. Why not abandon it and come out? Is there nothing else more worth your labour?' There is no one so stupid as to use this sort of language about mathematics. The mass of mathematical truth is obvious and imposing; its practical applications, the bridges and steam-engines and dynamos, obtrude themselves on the dullest imagination. The public does not need to be convinced that there is something in mathematics.

All this is in its way very comforting to mathematicians, but it is hardly possible for a genuine mathematician to be content with it. Any genuine mathematician must feel that it is not on these crude achievements that the real case for mathematics rests, that the popular reputation of mathematics is based largely on ignorance and confusion, and that there is room for a more rational defence. At any rate, I am disposed to try to make one. It should be a simpler task, at any rate, than Bradley's difficult apology.

I shall ask, then, why is it really worth while to make a serious study of mathematics? What is the proper justification of a mathematician's life? And my answers will be, for the most part, such as are to be expected from a mathematician: I think that it is worth while, that there is ample justification. But I should say at once that, in defending mathematics, I shall be defending myself, and that my apology is bound to be to some extent egotistical. I should not think it worth while to apologize for my subject if I regarded myself as one of its failures.

Some egotism of this sort is inevitable, and I do not feel that it really needs justification. Good work is not done by 'humble' men. It is one of the first duties of a professor, for example, in any subject, to exaggerate a little both the importance of his subject and his own importance in it. A man who is always asking 'Is what I do worth while?' and 'Am I the right person to do it?' will always be ineffective himself and a discouragement to others. He must shut his eyes a little and think a little more of his subject and himself than they deserve. This is not too difficult: it is harder not to make his subject and himself ridiculous by shutting his eyes too tightly.

10

A MATHEMATICIAN, like a painter or a poet, is a maker of patterns. If his patterns are more permanent than theirs, it is because they are made with *ideas*. A painter makes patterns with shapes and colours, a poet with words. A painting may embody an 'idea', but the idea is usually commonplace and unimportant. In poetry, ideas count for a good deal more; but, as Housman insisted, the importance of ideas in poetry is habitually exaggerated: 'I cannot satisfy myself that there are any such things as poetical ideas . . . . Poetry is not the thing said but a way of saying it.'

> Not all the water in the rough rude sea
> Can wash the balm from an anointed King.

Could lines be better, and could ideas be at once more trite and more false? The poverty of the ideas seems hardly to affect the beauty of the verbal pattern. A mathematician, on the other hand, has no material to work with but ideas, and so his patterns are likely to last longer, since ideas wear less with time than words.

The mathematician's patterns, like the painter's or the poet's, must be *beautiful*; the ideas, like the colours or the words, must fit together in a harmonious way. Beauty is the first test: there is no permanent place in the world for ugly mathematics. And here I must deal with a misconception which is still widespread (though probably much less so now than it was twenty years ago), what Whitehead has called the 'literary superstition' that love of and aesthetic appreciation of mathematics is 'a monomania confined to a few eccentrics in each generation.'

It would be difficult now to find an educated man quite insensitive to the aesthetic appeal of mathematics. It may be very hard to *define* mathematical beauty, but that is just as true of beauty of any kind—we may not know quite what we mean by a beautiful poem, but that does not prevent us from recognizing one when we read it. Even Professor Hogben, who is out to minimize at all costs the importance of the aesthetic element in mathematics, does not venture to deny its reality. 'There are, to be sure, individuals for whom mathematics exercises a coldly impersonal attraction . . . . The aesthetic appeal of mathematics may be very real for a chosen few.' But they are 'few', he suggests, and they feel 'coldly' (and are really rather ridiculous people, who live in silly little university towns sheltered from the fresh breezes of the wide open spaces). In this he is merely echoing Whitehead's 'literary superstition'.

The fact is that there are few more 'popular' subjects than mathematics. Most people have some appreciation of mathematics, just as most people can enjoy a pleasant tune; and there are probably more people really interested in mathematics than in music. Appearances may suggest the contrary, but there are easy explanations. Music can be used to stimulate mass emotion, while mathematics cannot; and musical incapacity is recognized (no doubt rightly) as mildly discreditable, whereas most people are so frightened of the name of mathematics that they are ready, quite unaffectedly, to exaggerate their own mathematical stupidity.

A very little reflection is enough to expose the absurdity of the 'literary superstition'. There are masses of chess-players in every civilized country—in Russia, almost the whole educated population; and every chess-player can recognize and appreciate a 'beautiful' game or problem. Yet a chess problem is *simply* an exercise in pure mathematics (a game not entirely, since psychology also plays a part), and everyone who calls a problem 'beautiful' is applauding mathematical beauty, even if it is beauty of a comparatively lowly kind. Chess problems are the hymn-tunes of mathematics.

We may learn the same lesson, at a lower level but for a wider public, from bridge, or descending further, from the puzzle columns of the popular newspapers. Nearly all their immense popularity is a tribute to the drawing power of rudimentary mathematics, and the better makers of puzzles, such as Dudeney or 'Caliban', use very little else. They know their business; what the public wants is a little intellectual 'kick', and nothing else has quite the kick of mathematics.

I might add that there is nothing in the world which pleases even famous men (and men who have used disparaging language about mathematics) quite so much as to discover, or rediscover, a genuine mathematical theorem. Herbert Spencer republished in his autobiography a theorem about circles which he proved when he was twenty (not knowing that it had been proved over two thousand years before by Plato). Professor Soddy is a more recent and a more striking example (but *his* theorem really is his own).*

---

*   See his letters on the 'Hexlet' in *Nature*, vols. 137-9 (1936-7).

G. PÓLYA

1887-1985

How to Solve It

1945

George Pólya was a mathematician born and educated in Hungary. He taught in Switzerland and then in the United States at Smith College, Brown, and Stanford Universities. His varied mathematical interests included series, number theory, combinatorics, and probability.

Pólya's legacy, however, results from his interest in the methods that people use to solve problems and how problem-solving should be taught and learned. One of his most useful and famous strategies is, "If you can't solve a problem, then there is an easier problem you can solve: find it." In the first of three books on problem solving, *How to Solve It*, "Part II: A Dialogue" is included in this volume. Pólya provides general heuristics for solving problems of all kinds, not simply mathematical ones. The book is for students and teachers alike and the preface to the first printing is a gem of personal, insightful thoughts on teaching and learning mathematics.

George Pólya is widely quoted and here are two favorites:

Mathematics in the primary schools has a good and narrow aim and that is pretty clear in the primary schools . . . . However, we have a higher aim. We wish to develop all the resources of the growing child. And the part that mathematics plays is mostly about thinking. Mathematics is a good school of thinking. But what is thinking? The thinking that you can learn in mathematics is, for instance, to handle abstractions. Mathematics is about numbers. Numbers are an abstraction. When we solve a practical problem, then from this practical problem we must first make an abstract problem . . . . But I think there is one point which is even more important. Mathematics, you see, is not a spectator sport. To understand mathematics means to be able to do mathematics. And what does it mean doing mathematics? In the first place it means to be able to solve mathematical problems.

Teaching is not a science; it is an art. If teaching were a science there would be a best way of teaching and everyone would have to teach like that. Since teaching is not a science, there is great latitude and much possibility for personal differences . . . . Let me tell you what my idea of teaching is. Perhaps the first point, which is widely accepted, is that teaching must be active, or rather active learning . . . . the main point in mathematics teaching is to develop the tactics of problem solving.[1]

---

[1] Pólya, G. H. *ca.*1969. "The Goals of Mathematical Education" videotaped lecture, transcribed by Thomas C. O'Brien. It first appeared in the *ComMuniCator*, the magazine of the California Mathematics Council.

Source:

O'Brien, Thomas C. 2001. "The Goals of Mathematical Education." Mathematically Sane. Retrieved February 22, 2007, from http://blk.mat.uni-bayreuth.de/aktuell/db/20/polya/polya.html.

Selection From:

Pólya, George. 1945. "Part II. How to Solve It: A Dialogue" In *How To Solve It: A New Aspect of Mathematical Method.* Princeton: Princeton University Press. 1973.

PART II. HOW TO SOLVE IT
A DIALOGUE

Getting Acquainted

Where should I start? Start from the statement of the problem.

What can I do? Visualize the problem as a whole as clearly and as vividly as you can. Do not concern yourself with details for the moment.

What can I gain by doing so? You should understand the problem, familiarize yourself with it, impress its purpose on your mind. The attention bestowed on the problem may also stimulate your memory and prepare for the recollection of relevant points.

Working for Better Understanding

Where should I start? Start again from the statement of the problem. Start when this statement is so clear to you and so well impressed on your mind that you may lose sight of it for a while without fear of losing it altogether.

What can I do? Isolate the principal parts of your problem. The hypothesis and the conclusion are the principal parts of a "problem to prove"; the unknown, the data, and the conditions are the principal parts of a "problem to find." Go through the principal parts of your problem, consider them one by one, consider them in turn, consider them in various combinations, relating each detail to other details and each to the whole of the problem.

What can I gain by doing so? You should prepare and clarify details which are likely to play a role afterwards.

Hunting for the Helpful Idea

Where should I start? Start from the consideration of the principal parts of your problem. Start when these principal parts are distinctly arranged and clearly conceived, thanks to your previous work, and when your memory seems responsive.

What can I do? Consider your problem from various sides and seek contacts with your formerly acquired knowledge.

Consider your problem from various sides. Emphasize different parts, examine different details, examine the same details repeatedly but in different ways, combine the details differently, approach them from different sides. Try to see some new meaning in each detail, some new interpretation of the whole.

Seek contacts with your formerly acquired knowledge. Try to think of what helped you in similar situations in the past. Try to recognize something familiar in what you examine, try to perceive something useful in what you recognize.

What could I perceive? A helpful idea, perhaps a decisive idea that shows you at a glance the way to the very end.

How can an idea be helpful? It shows you the whole of the way or a part of the way; it suggests to you more or less distinctly how you can proceed. Ideas are more or less complete. You are lucky if you have any idea at all.

What can I do with an incomplete idea? You should consider it. If it looks advantageous you should consider it longer. If it looks reliable you should ascertain how far it leads you, and reconsider the situation. The situation has changed, thanks to your helpful idea. Consider the new situation from various sides and seek contacts with your formerly acquired knowledge.

What can I gain by doing so again? You may be lucky and have another idea. Perhaps your next idea will lead you to the solution right away. Perhaps you need a few more helpful ideas after the next. Perhaps you will be led astray by some of your ideas. Nevertheless you should be grateful for all new ideas, also for the lesser ones, also for the hazy ones, also for the supplementary ideas adding some precision to a hazy one, or attempting the correction of a less fortunate one. Even if you do not have any appreciable new ideas for a while you should be grateful if your conception of the problem becomes more complete or more coherent, more homogeneous or better balanced.

Carrying Out the Plan

Where should I start? Start from the lucky idea that led you to the solution. Start when you feel sure of your grasp of the main connection and you feel confident that you can supply the minor details that may be wanting.

What can I do? Make your grasp quite secure. Carry through in detail all the algebraic or geometric operations which you have recognized previously as feasible. Convince yourself of the correctness of each step by formal reasoning, or by intuitive insight, or both ways if you can. If your problem is very complex you may distinguish "great" steps and "small" steps, each great step being composed of several small ones. Check first the great steps, and get down to the smaller ones afterwards.

What can I gain by doing so? A presentation of the solution each step of which is correct beyond doubt.

Looking Back

Where should I start? From the solution, complete and correct in each detail.

What can I do? Consider the solution from various sides and seek contacts with your formerly acquired knowledge.

Consider the details of the solution and try to make them as simple as you can; survey more extensive parts of the solution and try to make them shorter; try to see the whole solution at a glance. Try to modify to their advantage smaller or larger parts of the solution, try to improve the whole solution, to make it intuitive, to fit it into your formerly acquired knowledge as naturally as possible. Scrutinize the method that led you to the solution, try to see its point, and try to make use of it for other problems. Scrutinize the result and try to make use of it for other problems.

What can I gain by doing so? You may find a new and better solution, you may discover new and interesting facts. In any case, if you get into the habit of surveying and scrutinizing your solutions in this way, you will acquire some knowledge well ordered and ready to use, and you will develop your ability of solving problems.

LINUS PAULING

1901-1994

General Chemistry

1947

"No More War!"

1958

Linus Carl Pauling (1901-1994) is the only winner of two unshared Nobel Prizes—one for Chemistry and one for Peace. After receiving his PhD in chemistry and mathematical physics from the California Institute of Technology (Caltech) in 1925, he spent the next two years in Europe on a Guggenheim Fellowship. He joined the Caltech faculty in 1927. Along with his amazingly productive research projects, his duties at Caltech included teaching the freshman chemistry course and he was, by all accounts, a great lecturer. Through his laboratory demonstrations, which occasionally became pyrotechnical displays, he was able to explain difficult concepts in a charismatic and easy-to-understand way; he brought many of his new ideas about the chemical bond into the classroom. His *General Chemistry* (1947) has been used by generations of undergraduates, translated into 13 languages, and revised 3 times. His 1939 volume, *The Nature of the Chemical Bond,* is frequently cited as one of the most important scientific books of the 20th century. In 1954 he was awarded the Nobel Prize in Chemistry for his work in this area.

After WW II, Pauling became a pacifist and peace activist. Recognizing the dangers of nuclear weapons and radioactive fallout, he worked to prevent nuclear disaster. *No More War!* written with his wife, Ava Helen, presented the United Nations with a petition signed by more than 11,000 scientists calling for an end to nuclear-weapons testing. In 1963, on the same day that Pauling was awarded his second Nobel Prize, this time for Peace, the Partial Test Ban Treaty was signed by the U.S., Great Britain, and the U.S.S.R.—this agreement led to a moratorium on above-ground nuclear weapons testing. The Nobel selection committee noted that Pauling's continued successful efforts in the implementation of the ban, which could save innumerable people from cancer and genetic damage caused by ongoing testing.

In the late 1950s, Pauling turned his attention to human physiology and epidemiology to study the effects of smoking and other diseases. His interest became focused on the role of Vitamin C in maintaining health which led to the development of the field of orthomolecular medicine (meaning "right molecules in the right concentration"). His views are still controversial, but while the medical establishment has never fully endorsed his opinions, much of the general public did take Pauling's views to heart and he, in part, contributed to the enormous nutritional—supplements business in the United States.

Linus Pauling died in 1994 at 93. In a career that spanned seven decades, he wrote more than five-hundred papers and eleven books, and won every important prize awarded in his field.

Sources:

"Linus Pauling Biography." 2006. Linus Pauling Institute, Oregon State University. Retrieved February 3, 2007, from http://lpi.oregonstate.edu/lpbio/lpbio2.html.

"The Linus Pauling Papers." 2006. Profiles in Science, National Library of Medicine. Retrieved February 3, 2007, from http://profiles.nlm.nih.gov/MM/Views/Exhibit/narrative/biographical.html.

"Linus Pauling: The Nobel Prize in Chemistry 1954." 1954. Nobelprize.org. Retrieved February 3, 2007, from http://nobelprize.org/nobel_prizes/chemistry/laureates/1954/pauling-bio.html.

General Chemistry

1-1. Matter and Chemistry

The universe is composed of substances (forms of matter) and radiant energy. Chemistry is the science of substances—their structure, their properties, and the reactions that change them into other substances.

This definition of chemistry is both too narrow and too broad. It is too narrow because the chemist, in his study of substances, must also study radiant energy—light, X-rays, radiowaves—in its interaction with substances. He may be interested in the color of substances, which is produced by the absorption of light. Or he may be interested in the atomic structure of substances, as determined by the diffraction of X-rays (Chap. 3), or even by the absorption or emission of radiowaves by the substances.

On the other hand, the definition is too broad, in that almost all of science could be included within it. The astrophysicist is interested in the substances that are present in stars and other celestial bodies, or that are distributed, in very low concentration, through interstellar space. The nuclear physicist studies the substances that constitute the nuclei of atoms. The biologist is interested in the substances that are present in living organisms. The geologist is interested in the substances, called minerals, that make up the earth. It is, indeed, hard to draw a line between chemistry and other sciences.

The Structure of Matter. The properties of matter are most easily and clearly learned and understood when they are correlated with its structure, in terms of molecules, atoms, and still smaller particles. Structure has to do with the particles that go to make up all matter, and that through their interactions with one another give individuality and variety to matter. We shall launch upon the study of structure in the next chapter.

Kinds of Matter. As we look about us, we observe many different kinds of matter. We see a desk that is constructed mainly of wood, an organic material. The bracket holding up the arm of the desk is of iron; iron is a metal, and it is an elementary substance, one of the ninety-eight known chemical elements. The doorknob on the door is made of brass; brass also is a metal, but it is not an element: it is, instead, an alloy of the two elementary metals copper and zinc. The light fixtures are made of aluminum, copper, brass, tungsten, glass, fluorescent substances, mercury vapor, and several other materials. The student sitting at the desk is composed of matter of a great many different kinds.

With a microscope we can see the cells of plant and animal organisms, such as the red cells of the blood, which are about 0.001 cm in diameter (10^{-3} cm). With the electron microscope virus particles (virus molecules) 10^{-6} cm in diameter can be seen, and by means of the diffraction of X-rays and electron waves molecules and atoms approximately 10^{-8} cm in diameter can be studied. The physicists have, in fact, succeeded in investigating electrons, protons, neutrons, and other particles that are only about 10^{-12} cm in diameter; these are the smallest particles of matter that have as yet been discovered (Fig. 1-1).

The astronomers have obtained much information about matter outside of the earth. They have found that helium, sodium, calcium, hydrogen, and many other elements are present in the sun and other stars, and that ammonia, methane, and other substances are present in the atmospheres of the planets. They have found that matter is present in some stars in a very dense form: the density of one star, the companion of Sirius, is* 61,000 *g/cm³*, or about

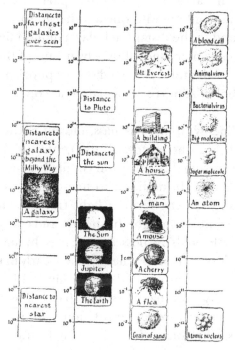

FIG. 1-1 *A diagram showing dimensions of objects, from 10^{-12} cm (the nucleus of an atom) to 10^{27} cm (the radius of the known universe).*

one ton per cubic inch. In interstellar space, on the other hand, the concentration of matter is very small: it has been estimated to be about one atom per cubic centimeter, which corresponds to about 10^{-23} g/cm³. We have knowledge about

---

*   The symbol *g/cm³* means grams per cubic centimeter. An alternative symbol is *g cm⁻³*. Either type of symbol may be used for simple combinations of units; for complex ones the symbols with negative exponents are preferred, because they are less liable to be misunderstood.

the nature of matter in the very distant nebulae, a billion light-years away—separated from us by what seems to be the radius of the universe, 10^{27} cm.

The chemist is interested in matter in all of its forms—in the minerals drugs, fuels, building materials, and living organisms on earth, in the minute particles 10^{-12} cm in diameter that combine with one another to form atoms and molecules, and in the distant nebulae a million light-years in diameter, that can be studied only by means of the light that reaches the earth and is collected onto a photographic plate by the paraboloidal mirror of a giant telescope.

Matter and Energy. Matter has mass, and any portion of matter on the earth is attracted toward the center of the earth by the force of gravity; this attraction is called the weight of the portion of matter. In addition to matter, the universe also contains energy, in the form of light (radiant energy). For many years scientists thought that matter and energy could be distinguished through the possession of mass by matter and the lack of possession of mass by energy. Then, early in the present century (1905), it was pointed out by Albert Einstein (born 1879) that energy also has mass, and that light is accordingly attracted by matter through gravitation. This was verified by astronomers, who found that a ray of light traveling from a distant star to the earth and passing close by the sun is bent toward the sun by its gravitational attraction. The observation of this phenomenon was made during a solar eclipse, when the image of the star could be seen close to the sun.

The amount of mass associated with a definite amount of energy is given by an important equation, the *Einstein equation:*

$$E=mc^2 \tag{1-1}$$

In this equation E is the amount of energy, m is the mass, and c is the velocity of light. The velocity of light, c, is one of the fundamental constants of nature;[*] its value is 2.9979 X 10^{10} cm/sec.

Until the present century it was also thought that matter could not be created or destroyed, but could only be converted from one form into another. In recent years it has, however, been found possible to convert matter into radiant energy, and to convert radiant energy into matter. The mass m of the matter obtained by the conversion of an amount E of radiant energy or convertible into this amount of radiant energy is given by the Einstein equation (1-1). Experimental verification of the Einstein equation has been obtained by the study of processes involving nuclei of atoms. The nature of these processes will be described in later chapters in this book.

The *units* of the quantities in the Einstein equation as written are those of the metric system (the centimeter-gram-second system). If m is given in g (grams) and c in cm/sec (centimeters per second), then the numerical value of mc^2 is the value of the energy E in ergs.

Until the present century scientists made use of a law of conservation of matter and a law of conservation of energy. These two conservation laws must now be combined into a single one, the **law of conservation of mass**, in which the mass to be conserved includes both the mass of matter in the system and the mass of energy in the system. However, for ordinary chemical reactions we may still make use of the "law" of conservation of matter—that matter cannot be created or destroyed, but only changed in form—recognizing that there is a limitation on the validity of this law: it is not to be applied if one of the processes involving the conversion of radiant energy into matter or matter into radiant energy takes place in the system under consideration.

Let us now consider again the two definitions given at the beginning of this chapter. We see that the statement that

---

[*]   The symbol c represents the velocity of light in a vacuum (empty space).

matter comprises all the substances of which the universe is composed is not really a definition until we have defined substances. Einstein's theory of relativity, which led to the relation between mass and energy, also provided a satisfactory definition of matter. According to the theory of relativity, matter comprises everything in the universe that has mass when it is standing still—this mass is called its *rest mass*. Additional energy *(kinetic energy)* is required to cause a portion of matter to move. The mass of the moving portion of matter is greater than the rest mass by an amount determined by the kinetic energy, according to Equation 1-1. According to the theory of relativity, it is impossible for any portion of matter to be accelerated to the speed of light. Light itself is considered to consist of bundles *(quanta)* of energy (these quanta of energy are also called *photons*), which can move only at the speed of light, and which have no rest mass. The nature of light will be discussed in detail in Chapter 8.

Units of the Metric System. The *mass* of an object is measured in terms of *grams* (g) or *kilograms* (kg), the kilogram being equal to 1,000 g. The kilogram is defined as the mass of a standard object made of a platinum-iridium alloy and kept in Paris. One pound is equal approximately to 454 g, and hence 1 kg is equal approximately to 2.2 lb. Note that it has become customary in recent years for the abbreviations of units in the metric system to be written without periods.

The metric unit of length is the *meter* (m), which is equal to about 39.37 inches. The meter is defined in terms of a standard meter, of platinum-iridium, kept in Paris. The *centimeter* (cm), which is 1/100 m, is about 0.4 inch, the inch being equal to 2.54 cm. The *millimeter* (mm) is 1/1,000 m or 1/10 cm.

The metric unit of volume is the *liter* (l), which is approximately 1.06 U.S. quarts. The *milliliter* (ml), equal to 1/1,000 l, is usually used as the unit of volume in the measurement of liquids in chemical work. The milliliter is defined as the volume occupied by 1 g of water at 3.98° C (the temperature at which its density is the greatest) and under a pressure of one atmosphere (that is, the normal pressure due to the weight of the air).

At the time when the metric system was set up, in 1799, it was intended that the milliliter be exactly equal to the cubic centimeter (cm^3). However, it was later found that the relation between the gram, as given by the prototype kilogram, and the centimeter, one one-hundredth of the distance between two engraved lines on a standard platinum-iridium bar (the prototype meter kept in Paris by the International Bureau of Weights and Measures), is such that the milliliter is not exactly equal to the cubic centimeter, but is, instead, equal to $1.000027\ cm^3$. It is obvious that the distinction between ml and cm^3 is ordinarily unimportant.

Units and Dimensions. We have stated above that the Einstein equation $E=mc^2$ is correct when m is expressed in grams, c in centimeters per second, and E in ergs. These quantities—grams, centimeters per second, and ergs—are called the *units* of m, c, E, respectively.

Other units might be used; for example, the mass might be expressed in pounds, the velocity of light in miles per second, or miles per hour, and the energy in calories or in some other energy unit. However, if other units were to be used for these quantities, it might be necessary to introduce a numerical factor into the equation. The units of the metric system were chosen in such a way as to make the numerical factors in the important equations of physics and chemistry as simple as possible; in this case, the case of the Einstein equation, the numerical factor has the value 1.

It is interesting to note that energy can be expressed in terms of units $g\ cm^2\ sec^{-2}$. There is another well-known equation of physics which involves the same sort of relationship. This is the equation expressing the kinetic energy of a moving particle in terms of its mass and its velocity, as given by Newton's laws of motion. The equation is

$$\text{kinetic energy} = \tfrac{1}{2}mv^2$$

The velocity, v, of a particle may be expressed, in the metric system, in units cm/sec—it is the quotient of the distance traversed by the particle and the time required for traversing this distance. Accordingly, the units of kinetic energy in the metric system are g cm^2 sec^{-2}. In the kinetic-energy equation there occurs the numerical factor ½.

No matter what units are used, an equation for a quantity of energy must always involve mass, length, and time in the following way, in which the square brackets are used to indicate the *dimensions* of the quantity within the brackets:

$$[energy] = [mass][length]^2[time]^{-2}$$

This equation says that the dimensions of energy are equal to the dimensions of mass times the dimensions of length squared times the dimensions of time with the exponent—2 (that is, divided by the dimensions of time squared).

It is good to develop the habit of checking the units and dimensions in the equations that you use.

1-2. *Kinds of Matter*

We shall first distinguish between objects and kinds of matter. An object, such as a human being, a table, a brass doorknob, may be made of one kind of matter or of several kinds of matter. The chemist is primarily interested not in the objects themselves, but in the kinds of matter of which they are composed. He is interested in the alloy brass, whether it is in a doorknob or in some other object; and, indeed, his interest is primarily in those properties of the material that are independent of the nature of the objects containing it.

The following sentences indicate the accepted scientific usage of words that designate different kinds of matter.

Materials. *The word* **material** *is used in referring to any kind of matter, whether homogeneous or heterogeneous.*

A *heterogeneous* material is a material that consists of parts with different properties. A *homogeneous* material has the same properties throughout.

Wood, with soft and hard rings alternating, is obviously a heterogeneous material, as is also granite, in which grains of three different species of matter (the minerals quartz, mica, and feldspar) can be seen.

Substances. *A* **substance** *is a homogeneous species of matter with reasonably definite chemical composition.*

Pure salt, pure sugar, pure iron, pure copper, pure sulfur, pure water, pure oxygen, and pure hydrogen are representative substances. On the other hand, a solution of sugar in water is not a substance, according to this definition: it is, to be sure, homogeneous, but it does not satisfy the second part of the above definition, inasmuch as its composition is not definite, but is widely variable, being determined by the amount of sugar that happens to have been dissolved in a given amount of water. Similarly, the gold of a gold ring or watchcase is not a pure substance, even though it is apparently homogeneous. It is an alloy of gold with other metals, usually copper, and it consists of a crystalline solution of copper in gold. The word *alloy* is used to refer to a metallic material containing two or more elements: some alloys are substances (inter-metallic compounds), but most of them are crystalline solutions or mixtures.

Sometimes (as in the first section of this chapter) the word "substance" is used in a broader sense, essentially as equivalent to material. Chemists usually restrict the use of the word in the way given by the definition above. The chemist's usage of the word substance may be indicated by using the phrase "pure substance."

Our definition is not precise, in that it says that a substance has reasonably definite chemical composition. Most materials that the chemist classifies as substances (pure substances) have definite chemical composition; for example, pure salt consists of the two elements sodium and chlorine in exactly the ratio of one atom of sodium to one atom of chlorine. Others, however, show a small range of variation of chemical composition; an example is the iron sulfide that is made by heating iron and sulfur together. This substance has a range in composition of a few percent. A discussion of substances with variable composition is given in Chapter 7.

Kinds of Definition. Definitions may be either precise or imprecise. The mathematician may define the words that he uses precisely; in his further discussion he then adheres rigorously to the defined meaning of each word. We have given some precise definitions above. One of them is the definition of the kilogram as the mass of a standard object, the prototype kilogram, that is kept in Paris. Similarly, the gram is defined rigorously and precisely as $1/1,000$ the mass of the kilogram.

On the other hand, the words that are used in describing nature, which is itself complex, may not be capable of precise definition. In giving a definition for such a word the effort is made to describe the accepted usage.

Mixtures and Solutions. A specimen of granite, in which grains of three different species of matter can be seen, is obviously a *mixture*. An emulsion of oil in water (a suspension of droplets of oil in water) is also a mixture. The heterogeneity of a piece of granite is obvious to the eye. The heterogeneity of an emulsion containing large drops of oil suspended in water is also obvious; the emulsion is clearly seen to be a mixture. But as the oil droplets in the emulsion are made smaller and smaller, it may become impossible to observe the heterogeneity of the material, and uncertainty may arise as to whether the material should be called a mixture or a solution.

An ordinary *solution* is homogeneous; it is not usually classified as a substance, however, because its composition is variable. A solution of liquids, such as alcohol and water, or of gases, such as oxygen and nitrogen (the principal constituents of air), may also be called a mixture. The word "mixture" may thus be used to refer to a homogeneous material that is not a pure substance or to a heterogeneous aggregate of two or more substances.

A homogeneous crystalline material is not necessarily a pure substance. Thus natural crystals of sulfur are sometimes deep-yellow or brown in color, instead of light-yellow. They contain some selenium, distributed at random throughout the crystals in place of some of the sulfur, the crystals being homogeneous, and with faces as well formed as those of pure sulfur. These crystals are a *crystalline solution* (or *solid solution*). The gold-copper alloy used in jewelry is another example of a crystalline solution. It is a homogeneous material, but its composition is variable.

Phases. A material system (that is, a limited part of the universe) may be described in terms of the *phases* constituting it. *A **phase** is a homogeneous part of a system, separated from other parts by physical boundaries.* For example, if a flask is partially full of water in whi   ch ice is floating, the system comprising the contents of the flask consists of three phases, the solid phase ice, the liquid phase water, and the gaseous phase air. A piece of malleable cast iron can be seen with a microscope to be a mixture of small grains of iron and particles of graphite (a form of carbon); it hence consists of two phases, iron and graphite (Fig. 1-2).

A phase in a system comprises all of the parts that have the same properties and composition. Thus, if there were several pieces of ice in the system discussed above, they would constitute not several phases, but only one phase, the ice phase.

Constituents and Components. Chemists use the words *constituent* and *component* in special ways.

*The **constituents** of a system are the various phases that constitute the system.*

A **set of components** *of a system is a set of substances (the minimum number of substances) from which the phases (constituents) of the system could be made.*

The constituents of the system discussed above are the three phases air, water, and ice. The components of the system may be taken to be either air and water, or air and ice, because the water phase and the ice phase can both be made from a single substance, water (or ice).* In this case the number of components is less than the number of phases. It may be greater; for example, a system consisting just of a solution of sugar in water is constituted of one phase, the solution, but it has two components, sugar and water.

1-3. *The Physical Properties of Substances*

Properties *of substances are their characteristic qualities.*

Sodium chloride, common salt, may be selected as an example of a substance. We have all seen this substance in what appear to be different forms—table salt, in fine grains; salt in the form of crystals a quarter of an inch in diameter, for use in regenerating water-softening minerals or with ice for freezing ice cream; and natural crystals of rock salt an inch or more across. Despite their obvious difference, all of these samples of salt have the same fundamental properties. In each case the crystals, small or large, are bounded by square or rectangular faces, of different sizes, but with each face always at right angles to each adjacent face. The possession of different properties in different directions—in particular the *formation of faces, edges, and corners*—is characteristic of crystals.

---

* In the above discussion air has been described as a component of the system. In discussing changes in state of the system in which the air behaves in the same way that nitrogen would behave, this would not lead to any difficulty, but in a rigorous treatment air might have to be described as involving several components (nitrogen, oxygen, argon, etc.).

The *cleavage* of the different crystals of salt is the same: when crushed, the crystals always break (cleave) along planes parallel to the original faces, producing smaller crystals similar to the larger ones. The different samples have the same salty *taste*. Their *solubility* is the same: at room temperature (18° C) 35.86 g of salt can be dissolved in 100 g of water. The *density* of the salt is the same, 2.163 g/cm^3.

Properties of this sort, which are not affected appreciably by the size of the sample or its state of subdivision, are called the *specific properties* of the substance represented by the samples.

There are other properties besides density and solubility that can be measured precisely and expressed in numbers. Such another property is the *melting point,* the temperature at which a crystalline substance melts to form a liquid. The *electric conductivity* and the *thermal conductivity* are similar properties. On the other hand, there are also interesting physical properties of a substance that are not so simple in nature. One such property is the *malleability* of a substance—the ease with which the substance can be hammered out into thin sheets. A related property is the *ductility*—the ease with which the substance can be drawn into a wire. *Hardness* is a similar property: we say that one substance is less hard than a second substance when it is scratched by the second substance, but this test provides only qualitative information about the hardness. A discussion of hardness is presented in Chapter 6.

The *color* of a substance is an important physical property. It is interesting to note that the apparent color of a substance depends upon its state of subdivision: the color becomes lighter as large particles are ground up into smaller ones.

It is customary to say that under the same external conditions all specimens of a particular substance have the same specific physical properties (density, hardness, color, melting point, crystalline form, etc.). Sometimes, however, the word "substance" is used in referring to a material without regard to its state of aggregation; for example, ice, liquid water, and water vapor may be referred to as the same substance. Moreover, a specimen containing crystals of rock salt and crystals of table salt may be called a mixture, even though the specimen may consist entirely of the one chemical substance sodium chloride. This lack of definiteness in usage seems to cause no confusion in practice.

The concept "substance" is, of course, an idealization; all actual substances are more or less impure. It is a useful concept, however, because we have learned through experiment that the properties of various specimens of impure substances with the same major component and different impurities are nearly the same if the impurities are present in only small amounts. These properties are accepted as the properties of the ideal substance.

1-4. *The Chemical Properties of Substances*

The **chemical properties** *of a substance are those properties that relate to its participation in chemical reactions.* **Chemical reactions** *are the processes that convert substances into other substances.*

Thus sodium chloride has the property of changing into a soft metal, sodium, and a greenish-yellow gas, chlorine, when it is decomposed by electrolysis. It also has the property, when it is dissolved in water, of producing a white precipitate when a solution of silver nitrate is added to it; and it has many other chemical properties. Iron has the property of combining readily with the oxygen in moist air, to form iron rust; whereas an alloy of iron with chromium and nickel (stainless steel) is found to resist this process of rusting. It is evident from this example that the chemical properties of materials are important in engineering.

Most substances have the power to enter into many chemical reactions. The study of these reactions constitutes a large part of the study of chemistry.

The properties of *taste* and *odor* are closely correlated with the chemical nature of substances, and are to be considered as chemical properties; the senses of smell and taste possessed by animals are the *chemical senses*. There is still complete lack of knowledge as to the way in which the molecules of tasty and odorous substances interact with the nerve endings in the mouth and nose to produce the sensations of taste and odor; this problem, like the problem of the molecular basis of the action of drugs, is one that awaits solution by the younger generation of chemists.

1-5. *The Scientific Method*

An important reason for studying science is to learn the scientific method of attack on a problem. The method may be valuable not only in the field of science but also in other fields—of business, of law, of government, of sociology, of international relations.

It is not possible to present a complete account of the scientific method in a few paragraphs. At this point there is given a partial account, which is amplified at the beginning of the following chapter, and in later chapters. Here I may say that the scientific method consists, in part, of the application of the principles of rigorous argument that are developed in mathematics and in logic, the deduction of sound conclusions from a set of accepted postulates. In a branch of mathematics the basic postulates are accepted as axioms, and the entire subject is then derived from these postulates. In science, and in other fields of human activity, the basic postulates (principles, laws) are not known, but must be discovered. The process of discovering these laws is called *induction*. The first step in applying the scientific method consists in finding some facts, by observation and experiment. In our science these are the facts of descriptive chemistry. The next step is the classification and correlation of many facts by one statement. Such a general statement, which includes within itself a number of facts, is called a *law*—sometimes a *law of nature*.

For example, when it was discovered, early in the nineteenth century, that water could be decomposed into hydrogen and oxygen by electrolysis (the action of an electric current), quantitative measurements of the amounts of hydrogen and oxygen were made. It was found in one experiment that 9 grams of water on electrolysis produced 1 gram of hydrogen and 8 grams of oxygen. This fact, for a particular specimen of water, was then amplified by additional facts, that the same amount of hydrogen, 1 gram, and the same amount of oxygen, 8 grams, were obtained by the electrolysis of 9 grams of water from other sources—rain water, sea water, water obtained by burning hydrogen in oxygen, etc. After many experiments of this sort had been made, all giving the same result, the facts were summarized in a law, that all samples of water give on electrolysis the same relative amounts of hydrogen and oxygen. When similar results were obtained for other chemical substances, this law was generalized into the law of constant composition (or law of definite proportions): in every pure sample of a given compound, the elements are present in the same proportion by weight.

It must be pointed out that the process of induction is never completely reliable. If one hundred analyses of water are made, by weighing the amounts of hydrogen and oxygen obtained by electrolysis of samples of water obtained from different sources, and the same proportion by weight of hydrogen to oxygen is found to within the limits of accuracy of the experiments, it would seem to be reasonably well justified to state that all samples of water have the same ratio of hydrogen to oxygen by weight. If a thousand analyses were made, with the same result, it would seem still more likely that this law is valid. However, if a single reliable analysis were then to be made which gave a different ratio, the law would have to be modified. It might turn out that the law is valid if the weighings of the gases are made with an accuracy of 0.1%, or 0.01%, but not if the weighings are made with still greater accuracy. This has, in fact, been found to be the case for water. In 1929 Professor William F. Giauque of the University of California at Berkeley discovered that there are three different kinds of oxygen atoms, with different masses (these atoms are called isotopes; see Chap. 4), and shortly thereafter Professor Harold C. Urey discovered that there are two kinds of hydrogen atoms, with different masses. Water consisting of molecules made with these different kinds of hydrogen atoms and oxygen atoms contains hydrogen and oxygen in different ratios by weight, and it has been found that the composition by weight of pure water from different natural sources is, in fact, slightly different. It has accordingly become necessary to revise the law of constant composition in such a way as to take into account the existence of these isotopic forms of atoms. The way in which this is done is described in Chapter 4.

One important way in which progress has been made in science is through a *process of successive approximations.* Some measurements are made with a certain precision, such as the measurements of the compositions of substances with accuracy of 1%, and a rough law is formulated that encompasses all of these measurements. It may then happen that, when more precise measurements are made, deviations from the first law are found to exist. A second, more refined but more complicated law may then be formulated to include these deviations. This procedure may have been carried out several times in the course of the formulation of a law of nature in its now accepted state.

It is wise to remember that a law obtained by the process of induction may at any time be found to have limited validity. Conclusions that are reached from such a law by the process of deduction should be recognized as having a probability of being correct that is determined by the probability that the original law is correct.

The application of the scientific method does not consist solely of the routine use of logical rules and procedures. Often a generalization that encompasses many facts has escaped notice until a scientist with unusual insight has discovered it. Intuition and imagination play their part in the scientific method.

As more and more people gain a sound understanding of the nature of the scientific method and learn to apply it in the solution of the problems of everyday life, we may hope for an improvement in the social, political, and international affairs of the world. Technical progress represents one way in which the world can be improved through science. The other way is through the social progress that results from application of the scientific method—through the development of "moral science"; and I believe that the study of science, the learning of the scientific method by all people, will ultimately help the people of the world in the solution of our great social and political problems.

No More War!

1

The End of War

I believe that there will never again be a great world war, if only the people of the United States and of the rest of the world can be informed in time about the present world situation. I believe that there will never be a war in which the terrible nuclear weapons—atom bombs, hydrogen bombs, superbombs—are used. I believe that the development of these terrible weapons forces us to move into a new period in the history of the world, a period of peace and reason, when world problems are not solved by war or by force, but are solved by the application of man's power of reason, in a way that does justice to all nations and that benefits all people.

I believe that this is what the future holds for the world, but I am sure that it is not going to be easy for the world to achieve this future. We have to work to prevent the catastrophe of a cataclysmic nuclear war, and to find the ways in which world problems can be solved by peaceful and rational methods.

In the past, disputes between groups of human beings have often been settled by war. At first the wars were fought with stones and clubs as weapons, then with spears and swords, and then with bows and arrows. During the last few hundred years they have been fought with guns, and recently with great bombs dropped from airplanes—blockbusters containing one ton or even ten tons of TNT.

There may in the past have been times when war was a cruel but effective application of the democratic process, when force was on the side of justice. Wars fought with simple weapons were often won by the side with the greater number of warriors.

Now war is different. A great mass of people without nuclear weapons, without airplanes, without ballistic missiles cannot fight successfully against a small group controlling these modern means of waging war.

The American people were successful in their revolt against Great Britain because modern weapons had not yet been developed when the American Revolution broke out.

It is hard for anybody to understand how great a change has taken place in the nature of the world during the past century or two, and especially during the last fifty years. The world has been changed through the discoveries made by scientists. Everything has been changed by these discoveries—the food we eat, the clothes we wear, the methods of controlling disease, the methods of transportation and communication, the conduct of international affairs, the ways of waging war—all are different now from what they were a few decades ago.

Never again will there be the world of William Shakespeare, the world of Benjamin Franklin, the world of Queen Victoria, the world of Woodrow Wilson.

The scientific discoveries that have changed the world are manifold. I think that the greatest of all scientific discoveries, the greatest discovery that has been made since the discovery of the controlled use of fire by prehistoric man, was the discovery of the ways in which the immense stores of energy that are locked up in the nuclei of atoms can be released.

Many scientists contributed to this discovery. Among them we may mention some of the great ones—Pierre and Marie Curie, Albert Einstein, Ernest Rutherford, Niels Bohr, Ernest

Lawrence, Frederic and Irene Joliot-Curie, Otto Hahn, Enrico Fermi.

This discovery, by providing power in essentially unlimited quantities for the future world, should lead through its peaceful applications to a great increase in the standards of living of people all over the world.

It is this discovery also that has changed the nature of war in an astounding way.

The Second World War, like earlier wars, was fought with molecular explosives. Trinitrotoluene, TNT, is the one that was most used. TNT can be manufactured by the reaction of nitric and sulfuric acids with the hydrocarbon toluene, which can be obtained from petroleum. It is not very expensive—somewhere around 25 cents a pound. One pound of TNT can do a lot of damage; about the same as one stick of dynamite. It can demolish a small house and kill several people. A one-ton blockbuster, a bomb containing 2000 pounds of TNT, can demolish a large building and may kill a hundred people or more.

During the Second World War many shells and bombs containing TNT and other explosives were fired at or dropped on cities and other targets in the warring countries. The total amount of explosives used in the whole of the Second World War came to about three million tons of TNT.

In this book we shall often refer to the explosive energy of one million tons of TNT. We may call this one megaton. We should remember that three megatons is the equivalent of all the explosives used in the whole of the Second World War.

At the beginning of the Second World War the discoveries in the field of nuclear physics had just reached such a point as to cause a number of scientists to recognize that it might be possible to manufacture immensely powerful bombs involving nuclear reactions, and also to make nuclear power plants.

In March 1939 Enrico Fermi had a conference with representatives of the Navy Department, with the outcome that the Navy expressed interest and asked to be kept informed.

In July 1939 Leo Szilard and Eugene Wigner conferred with Einstein, and a little later Einstein, Szilard, and Wigner discussed the matter with Alexander Sachs. Sachs, supported by a letter from Einstein, then explained the situation to President Roosevelt, who appointed a committee, the "Advisory Committee on Uranium," to look into the problem. In December 1941, after receipt of information about progress made independently by the British, the atomic bomb project was initiated, and in a few years the thousands of American, British, French, and other scientists who had been brought together succeeded in developing atomic bombs of two types, the Hiroshima type (fission of uranium-235) and the Nagasaki type (fission of plutonium-239).

At 8:15 A.M. of 6 August 1945 the first atomic bomb used in war was dropped on Hiroshima, a city of about 450,000 people in the western part of Japan. It was Monday, and the people were setting out on their way to work. The bomb was dropped from an American B-29 flying at about 24,000 feet. The B-29 flew away at full speed as the bomb, supported by a parachute, slowly descended. It exploded above the central part of the city, at a height of about 2200 feet. Within a few seconds the blast from the bomb destroyed 60 percent of the city. Many thousands of people were killed by the blast itself or crushed by falling buildings. Many others were killed by burns, caused by the great amount of radiation from the fireball of the bomb, which had a surface temperature greater than that of the sun. Many thousands of people received exposure to ionizing radiation that caused them to die of radiation sickness in a few days.

About 100,000 people were killed by the bomb in Hiroshima and about another 100,000 seriously injured.

On 9 August 1945, three days after the Hiroshima bomb had been dropped, a second atomic bomb was dropped on Japan. It exploded over Nagasaki, a city of about 300,000 population in southern Japan, on the island of Kyushu. This bomb destroyed a large part of the city, killed about 70,000 people, and seriously injured about another 70,000 people.

The Nagasaki and Hiroshima bombs had explosive energy somewhere between 15,000 and 20,000 tons of TNT. Each of them was accordingly about 15,000 or 20,000 times more powerful than a one-ton blockbuster. Each was about 1000 times as powerful as the greatest of the great bombs with conventional explosives used in the Second World War. Each of them killed more than ten thousand times as many people as were killed by the average blockbuster of the Second World War.

It was clear that war had entered into a new period—the period of atomic bombs.

Within a few days Japan had surrendered (14 August 1945). It was evident that no nation could continue to fight against an enemy nation possessing these terrible weapons. I think that the surrender would have come if the bombs had been dropped in the open country, rather than on the cities of Hiroshima and Nagasaki—the power of the great new weapons would have been demonstrated nearly as clearly in this way, without the great loss of life by the women, children and other non-combatants of the two cities.

In the years following the end of the Second World War many people pointed out that a war waged with atom bombs would produce a devastation incomparably greater than that of the Second World War, would kill hundreds of millions of human beings.

Albert Einstein in 1946 said:

> A new type of thinking is essential if mankind is to survive and move to higher levels. Today the atomic bomb has altered profoundly the nature of the world as we know it and the human race consequently finds itself in a new habitat to which it must adapt its thinking. Modern war, the bomb, and other discoveries present us with revolutional circumstances. Never before was it possible for one nation to make war on another without sending armies across borders. Now with rockets

and atomic bombs no center of population on the earth's surface is secure from surprise destruction in a single attack. Should one rocket with atomic warhead strike Minneapolis, that city would look almost exactly like Nagasaki. Rifle bullets kill men, but atomic bombs kill cities. A tank is a defense against a bullet, but there is no defense in science against a weapon which can destroy civilization.

Our defense is not in armaments, nor in science, nor in going underground. Our defense is in law and order.

Henceforth every nation's foreign policy must be judged at every point by one consideration: Does it lead us to a world of law and order or does it lead us back toward anarchy and death? I do not believe that we can prepare for war and at the same time prepare for a world community. When humanity holds in its hand the weapon with which it can commit suicide, I believe that to put more power into the gun is to increase the probability of disaster.

The world had indeed entered into a new stage of development in 1945, when the atom bombs were dropped on Hiroshima and Nagasaki and the United States began to build up its stockpile of thousands of these terribly destructive weapons, each one capable of destroying a medium-sized city and killing a hundred thousand people.

Albert Einstein was justified in his great apprehension, his fear that the possession of weapons of destruction a thousand times more powerful than the greatest that were ever used before would lead to catastrophe for the world.

And now we have the H-bomb!

Now the United States, the U.S.S.R., and Great Britain have stockpiles of hydrogen bombs and superbombs that are one thousand

*times more powerful still than the atomic bombs that were dropped on
Hiroshima and Nagasaki!*

In 1945 the world changed from the period of TNT
blockbusters, with war as in the Second World War, when one
large bomb could kill ten people or a hundred people, into
its second period, the period of the great atomic bombs, each
capable of killing one hundred thousand people. In 1952 the
world moved into the *third* period, when the bombs became
not just one thousand or ten thousand times more powerful
than the blockbusters, but *one million* or *ten million* times as
powerful—*one thousand times more powerful than the Hiroshima
and Nagasaki bombs.*

If a war were to break out today, it is probable that one
bomb would explode over New York, and kill ten million
people. One bomb would explode over London, and kill
ten million people. One bomb would explode over Moscow,
and kill six million people. One bomb would explode over
Leningrad, and kill three million people. One bomb would
explode over Chicago, and kill four million people. One bomb
would explode over Los Angeles, and kill three million people.
And these cities themselves would be smashed flat, over an
area ten miles or twenty miles in diameter. The cities and the
regions around them would be rendered uninhabitable for
years by the deposited radioactivity. The initial attacks in such
a war would kill 83 million Americans and seriously injure
another 25 million.

The bomb that could destroy the greatest city in the world
and kill ten million people is not something imaginary. Bombs
of this sort—hydrogen bombs and superbombs—have been
made and have been exploded. Bombs have been tested that
have an explosive power as great as 15 megatons—an explosive
power equivalent to 15 million tons of TNT, 15 million one-ton
blockbusters.

Each one of these bombs is one thousand times more
powerful than the Hiroshima bomb or the Nagasaki bomb.

Each one of them has an explosive energy five times as great as that of all of the bombs used in the Second World War.

Albert Einstein was apprehensive, fearful, in 1946, when the world had passed into the period of atomic bombs, little nuclear bombs such that each bomb equals one hundred thousand people killed. How much more reason do we have now for fear about the future of the world, now that we have moved into the period of great nuclear weapons, with each bomb equal to as many as ten million people killed!

We are truly forced into abandoning war as the method of solution of world problems, the method of resolution of disputes among nations.

President Eisenhower and other great national leaders have stated that the stockpiles of these terrible nuclear weapons are deterrents—that they will serve as deterrents against aggression, deterrents against war. There is little doubt that the nuclear weapons have been effective in preventing the outbreak of great wars during recent years.

But disputes between nations still need to be settled. In the past, world problems have often been settled by war, sometimes in such a way as to correspond to justice and sometimes to injustice. If the nuclear stockpiles continue to serve effectively as deterrents and to prevent war from breaking out, there still remains the task of solving the great world problems.

It is evident that these problems now need to be settled by the processes of negotiation, arbitration, the formulation and application of a sound system of international law. *We need to begin now to make international agreements.*

I am not alone in this belief. I think that the great majority of the scientists in the world and the great majority of the people in the world believe that there must be no more war, that the great stockpiles of terribly destructive nuclear weapons must not be used, that the time has come for morality and justice to take their proper place of prime importance in the conduct of world affairs, that world problems should be settled by international agreements and the application of international law.

On 13 January 1958 I presented to the United Nations a petition to which 9235 scientists, of many countries in the world, had subscribed. In this petition we urged that immediate action be taken to effect an international agreement to stop the testing of all nuclear weapons. We pointed out that if testing continues and the possession of the nuclear weapons spreads to additional governments there will be greatly increased danger of outbreak of a cataclysmic nuclear war through the reckless action of some irresponsible national leader. We mentioned also the damage that is being done to the health of human beings all over the world and to the pool of human germ plasm that determines the nature of future generations by the bomb tests, which spread radioactive elements over every part of the world. We mentioned that an international agreement to stop the testing of nuclear bombs now could serve as the first step toward averting the possibility of a nuclear war that would be a catastrophe for all humanity. Our proposal is that we begin now the new era of peace and international law, by making an international agreement to stop the bomb tests.

It is these matters that will be discussed in some detail in the later chapters of this book. Also, in the final chapter a proposal is made about how peace is to be achieved in the world, how the great world problems are to be solved without recourse to war, war that would now lead to catastrophe, to world suicide.

It is proposed that the great world problems be solved in the way that other problems are now solved—by working hard to find their solution—by carrying on *research for peace*. It is proposed that there be set up a great research organization, the World Peace Research Organization, within the structure of the United Nations. This organization should include many scientists, representing all fields of science, and many other specialists, in all fields of knowledge. They would attack world problems by imaginative and original methods, and would work steadily in this attack, year after year.

The time has now come for man's intellect to win out over the brutality, the insanity of war.

2

The Nature of Nuclear Weapons

Before we discuss nuclear weapons, let us have a look at one of the old-fashioned explosives, TNT.

TNT is the substance trinitrotoluene. It is made by pouring toluene into a mixture of nitric acid and sulfuric acid. Toluene is a liquid something like benzene. It is found in small quantity in some kinds of petroleum and is easily made from the other constituents of petroleum or natural gas.

TNT is a yellow solid crystalline substance with the chemical formula $C_7H_5N_3O_6$. The crystals consist of molecules, each of which is made of seven carbon atoms, five hydrogen atoms, three nitrogen atoms, and six oxygen atoms, which are arranged in space in the way shown in Figure 1. The drawing shown in Figure 1 represents a magnification of 100 million-fold.

This is a great amount of magnification. If we had one pound of TNT in the form of a ball it would be about three inches in diameter. Now if this ball were to be magnified 100 million times, so that each of the molecules of TNT in it would have the size shown in the figure, the ball itself would have become over 5000 miles in diameter, and hence nearly as large as the earth.

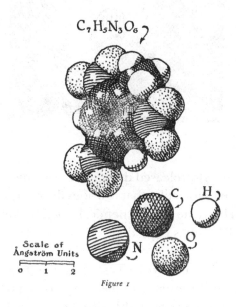

Figure 1

When a mass of TNT is detonated by means of a sharp blow, such as that produced by a detonator, the atoms in the molecule separate from one another and unite to form new molecules. These are small molecules, such as shown in Figure 2, where molecules of nitrogen, carbon monoxide, and hydrogen are represented. The bonds holding the atoms together in these molecules are stronger than the bonds in the molecule of TNT, and the extra energy that is stored in the bonds of the TNT molecules is released in the explosion. The amount of this energy is such that the products of the explosion reach a temperature of nearly 4000°C.

Figure 2

It is the energy released during the rearrangement of the atoms into new molecules that provides the energy of the TNT explosion. This is the sort of chemical reaction that takes place when a TNT bomb is exploded.

Other molecular explosives, such as nitroglycerin, get their energy from similar chemical reactions. The amount of energy that can be provided by such a molecular explosive is never very much greater than that which is provided by TNT. There is no chance that superbombs can ever be built of molecular explosives.

A molecule such as TNT is made of electrons and atomic nuclei. The electrons and the atomic nuclei are extremely small, over 10,000 times smaller in diameter than the atoms themselves, represented in Figures 1 and 2. Each carbon atom in the TNT molecule has one nucleus. Each atom of hydrogen, nitrogen, and oxygen has one nucleus. These nuclei are not changed at all in the explosion of TNT, nor are the electrons changed; there is only a rearrangement of the atoms during this reaction.

The Hiroshima Bomb

Now let us consider an atomic explosive; for example, uranium-235, which was used in making the Hiroshima bomb. Uranium-235 is a hard, heavy, white metal. It is nearly as dense as gold: a ball of it three inches in diameter weighs about 10 pounds. Such a ball might constitute the explosive charge of an atomic bomb similar to the Hiroshima bomb.

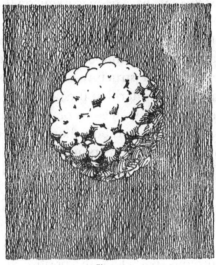

Figure 3

The uranium-235 metal consists of uranium atoms, each of which is a nucleus of uranium-235 surrounded by 92 electrons. When the bomb is exploded it is the uranium-235 nuclei that react.

In order to make a drawing of the nucleus of uranium-235 we shall need to use a scale of magnification about 50,000 times that used for the molecule of TNT. Such a drawing is shown in Figure 3. It represents a magnification of five million million times; that is, 5,000,000,000,000.

This magnification is truly immense. If we had a ball of uranium-235 weighing 10 pounds, and could enlarge it until the nucleus of each uranium atom had the size shown by the drawing, the ball itself would have become five hundred million miles in diameter. Its diameter would be greater than the distance from the earth to the sun.

The drawing shown in Figure 3 is a somewhat imaginative one. We know a great deal about the molecule of TNT; all of the interatomic distances are known to within about 1 percent, and the drawing shown in Figure 1 is a thoroughly trustworthy one. However, even though the hundreds of physicists who during the last 25 years have been working on the problem of the structure of the atomic nuclei have learned a great deal, they have not yet succeeded in obtaining the sort of detailed information about the structure of the atomic nuclei that has been obtained about the structure of molecules. This is not surprising when we consider that the nuclei are 50,000 times smaller than the molecules—it is one of the marvels of modern science that so much is known about the structure of molecules, and we may have to wait a few years more before similar precise information has been obtained about the nuclei.

The little particles that are represented as forming the nucleus of uranium-235 in Figure 3 are the nucleons. The nucleons are protons and neutrons. It is customary to describe a nucleus of uranium-235 as consisting of 235 nucleons of which 92 are protons (particles with a positive charge) and 143 are neutrons (particles with no electric charge). The protons and the neutrons have nearly the same mass; each has a mass approximately $1/_{235}$ of the mass of the uranium-235 nucleus.

The nucleons are held together by forces that in a general sort of way are analogous to, but far stronger than, the forces between atoms that constitute the chemical bonds in the TNT molecule. Under certain circumstances the nucleons may be rearranged, just as the atoms in the molecule of TNT are rearranged during the explosion of TNT.

In the explosion of an atomic bomb this rearrangement occurs after the nucleus has collided with a free neutron.

Reactions of this sort take place all the time in every piece of uranium metal and in every sample of any compound of uranium. There are a few neutrons running around everywhere. A neutron may result, for example, from the collision of a cosmic ray, coming in from outer space, with an atom. If this neutron collides with a nucleus of uranium-235 it may be absorbed by the nucleus, which then contains 236 nucleons, rather than 235—it has become a nucleus of uranium-236. The nucleus of uranium-236 is unstable, in such a way that it spontaneously rearranges, decomposes, as shown in Figure 4. The nucleus splits into two half-sized nuclei, which may be called daughter nuclei. Not all of the nucleons remain in the two daughter nuclei; instead two or three escape as free neutrons. Sometimes the splitting of the uranium-236 nucleus occurs in such a way as to give three daughter nuclei, rather than two, plus a few free neutrons.

In this reaction, which is called the *fission* of the nucleus, a tremendous amount of energy is liberated. The amount of energy liberated by the fission of one uranium nucleus is about ten million times as great as the amount liberated in the decomposition of one molecule of TNT. The molecule of TNT has about the same weight as the atom of uranium; accordingly the explosive uranium-235 is, on a weight basis, about ten million times as powerful as TNT.

Figure 4

A small piece of uranium-235, a few pounds of the metal, will not explode under ordinary conditions. Every once in a while a stray neutron will bump into one of the uranium nuclei and cause it to undergo fission. Two or three neutrons may be liberated in the process of fission, but in general they escape to the surrounding world. Occasionally one of them may hit another uranium nucleus and cause it to undergo fission, too, but the small piece of uranium-235 does not explode.

However, a large piece of uranium-235 is very dangerous—it will explode spontaneously. The size of the piece of the metal that explodes spontaneously is called the *critical mass*. The process that is involved in its spontaneous explosion is called a *chain reaction*.

We may describe the way in which the chain reaction operates. A stray neutron hits a uranium nucleus, perhaps near the center of the piece of uranium-235, and causes it to undergo fission. Let us assume that two neutrons are liberated in the process of fission. If the piece of uranium-235 is a big one, each of these two neutrons will bump into another uranium nucleus, and there will be two more nuclei that undergo fission, and four neutrons will be liberated. These four neutrons will then cause four uranium nuclei to undergo fission, liberating eight neutrons. As this chain reaction continues there will be sixteen, thirty-two, sixty-four, and so on, nuclei undergoing fission, and within a millionth of a second most of the uranium-235 nuclei will have been caused to decompose. This is the process of explosion of the fissionable material of the atomic bombs.

I do not know exactly how the Hiroshima atomic bomb was built. Perhaps it consisted of two little pieces of uranium-235, each weighing perhaps five pounds, which were held a few inches away from one another in the bomb. There might have been then a couple of pieces of TNT or some similar ordinary explosive on the far sides of the two pieces of uranium-235, such that when they were detonated they would push the two pieces of uranium-235 together, producing a mass greater than the critical mass. Then the chain reaction would begin, and within a millionth of a second the atomic bomb would have exploded—and then within a few seconds many of the hundred thousand victims would be dead.

There is another possibility. Perhaps a mass of, say, ten pounds of uranium-235 is stable under ordinary circumstances, but becomes unstable, so as to explode by the chain reaction, when it is compressed into a smaller volume. The bomb might thus have been made of a ball of uranium-235 about three inches in diameter (perhaps with a central cavity) and weighing about ten pounds, surrounded by some TNT or other molecular explosive with detonators arranged in such a way that when the explosion of the ordinary explosive occurred the ball of uranium-235 would be subjected to great pressure, and be compressed down to two inches in diameter, and this small ball of compressed uranium-235 would then undergo the chain reaction.

The Nagasaki Bomb

The Nagasaki bomb was closely similar to the Hiroshima bomb, but used plutonium-239 as the explosive, in place of uranium-235.

The uranium-235 of the Hiroshima bomb had been separated from natural uranium metal by a physical process. Natural uranium metal consists almost entirely of the isotope uranium-238; only about 0.7 percent is uranium-235. Uranium-238 cannot be used by itself to make an atomic bomb, nor can natural uranium, the mixture of uranium-238 and uranium-235, be used. Some great plants were built in Oak Ridge to carry out the difficult process of separating the uranium-235 from natural uranium.

A second way of obtaining fissionable material was put into operation in a great plutonium plant in Hanford, Washington. Rods of uranium metal, ordinary uranium, are piled together with rods of graphite, which serves to slow down the neutrons, which are ejected at high speed from the fissioning nuclei. The neutrons that are produced by the fission of uranium-235 are captured by the nuclei of uranium-238, which are then converted into nuclei of uranium-239. The nuclei of uranium-239 decompose spontaneously, with the emission of an electron, to produce nuclei of neptunium-239, and these nuclei soon decompose spontaneously in the same way with the production of nuclei of plutonium-239. The nucleus of plutonium-239 contains 94 protons and 145 neutrons.

Plutonium has chemical properties greatly different from those of uranium, and the plutonium made in this way can be easily separated from the uranium by chemical methods. These methods are so much simpler than the physical methods needed to separate uranium-235 from uranium-238 as to give plutonium an advantage as a fissionable material. In addition, its properties as a nuclear explosive are somewhat superior to those of uranium-235.

Plutonium-239 is a rather stable substance; under ordinary conditions only about half of it has decomposed after 25,000 years. But it is fissionable in the same way as uranium-235. If a mass of plutonium-239 metal is compressed or brought together in such a way as to permit the chain reaction to go on, the plutonium-239 combines with neutrons and undergoes fission to half-size or smaller nuclei, with the liberation of additional free neutrons, and in a small fraction of a second the nuclear explosion has taken place. It was perhaps ten pounds of plutonium-239 made in Hanford that constituted the explosive of the Nagasaki bomb.

It is possible to carry out the explosion of a smaller amount of uranium-235 or plutonium-239; atomic bombs somewhat smaller than the Hiroshima bomb and the Nagasaki bomb have been tested. I do not know just how small the smallest ones are—perhaps about the equivalent of 1000 tons of TNT, involving the fission of an ounce or two of the fissionable metal. A bomb exploded in an underground test in Nevada on 19 September 1957 was stated by a representative of the

Atomic Energy Commission to be a 1.7-kiloton bomb—the equivalent of only 1700 tons of TNT.

Also, it is possible to make still larger bombs with uranium-235 or plutonium-239, as large as about 100 kilotons equivalent of TNT. There may be technical difficulties in making these bombs much larger than this.

But the problem of making extremely powerful bombs has been solved. The solution that has been found will be described in the following two sections.

Hydrogen Bombs

Energy is liberated by the decomposition of molecules, such as those of TNT, as we have mentioned at the beginning of this chapter. Also, energy is liberated in some chemical reactions that involve the formation of larger molecules, rather than their decomposition. For example, hydrogen and oxygen burn together to produce water. The hydrogen-oxygen flame is a very intense one, giving a white heat. In this chemical reaction the molecules that react contain only two atoms apiece, and the molecules that are produced, those of water, contain three atoms apiece.

In an analogous way there occur not only nuclear reactions in which large nuclei decompose into smaller ones, by fission, but also nuclear reactions in which larger nuclei are formed from smaller ones. This process is called *fusion*. Fusion and fission are the two kinds of nuclear reactions that are used in nuclear weapons.

Perhaps the most important fusion reaction that takes place is the reaction of hydrogen nuclei (protons) to form helium nuclei (made of four nucleons—two protons and two neutrons). This reaction takes place, in a rather complicated way, in the interior of the sun. It is this fusion process that provides the energy that is radiated by the sun, and that has kept the sun hot for billions of years. The fusion reaction of hydrogen nuclei to form helium nuclei liberates about 50 million times as much energy as is liberated by the reaction of the same weight of hydrogen molecules and oxygen molecules to form water molecules.

There are various difficulties in making hydrogen bombs—bombs that use nuclear fusion for the major part of their explosive energy. However, these difficulties were overcome at about the same time, five years ago, by the American bomb-makers and the Russian bomb-makers; and about one year ago (1957) the British bomb-makers, too, exploded some hydrogen bombs.

The atomic nuclei have a positive electric charge. Two particles with a positive electric charge repel one another, and hence under ordinary conditions two protons, for example, will not get close enough to one another to undergo nuclear reaction. At extremely high temperatures, of the order of 50 million degrees centigrade, the nuclei have so much energy of translational motion—they are moving so rapidly around—that they are able to collide with one another effectively and to undergo reaction. Fusion reactions can accordingly be made to take place by providing a way of heating the reaction materials to about 50 million degrees.

So far the success of the hydrogen-bomb program has depended upon a method of obtaining temperatures of about 50 million degrees that has been at hand since 1945: when an atomic bomb of the Hiroshima type or the Nagasaki type explodes it produces a body of hot gas with temperature about 50 million degrees.

A hydrogen bomb is made by starting with an ordinary atomic bomb, in which the fission process of uranium-235 or plutonium-239 takes place. This atomic bomb acts as the detonator for the fusion process. It heats the hydrogen nuclei or other light nuclei that are to undergo fusion to a very high temperature, so that some of them collide with one another and react, liberating great amounts of fusion energy. The temperature may become even higher as a result of the fusion process, and more of the fusionable material may then react.

If a few pounds of the fissionable material is used as a detonator, and it is surrounded by several hundred pounds or even a ton or two of fusionable material, a hydrogen bomb can be made. Such a bomb is not much greater in weight than a one-ton TNT blockbuster of the good old days of molecular explosives, but it has an explosive energy 10 million or 20 million times that of the one-ton blockbuster, one thousand times as great as that of a Hiroshima or Nagasaki bomb.

There may be no limit to the size of such a hydrogen bomb. So far as I know, the biggest ones that have been tested have been in the range of five to ten megatons—that is, each one of them equivalent to twice or three times all of the bombs exploded in the Second World War. Perhaps the principal reason for not making them any bigger is that there is no target in the world that requires a bomb much bigger than ten megatons.

Ordinary hydrogen, with a proton as the nucleus of its atoms, is not the principal material used in hydrogen bombs. There are two heavier isotopes of hydrogen. One of these is deuterium, which has a nucleus made of two nucleons, a proton and a neutron. This heavy hydrogen was discovered by Professor Harold C. Urey and his collaborators about twenty-five years ago. Another isotope of hydrogen is tritium. Tritium has mass three—its nucleus consists of a proton and two neutrons. The tritium nucleus is unstable, and tritium does not occur in significant amount in nature. It can, however, be made by the use of neutrons from a nuclear reactor.

A hydrogen bomb probably could be constructed with the use of a mixture of deuterium and tritium, plus, of course, the ordinary atomic bomb as its detonator. However, elementary deuterium and tritium, like ordinary hydrogen, are gases at room temperature and atmospheric pressure, and it would be quite inconvenient to have to liquefy these gases and to keep the hydrogen bomb at an extremely low temperature, close to the absolute zero, until it was exploded. A nuclear device of this sort, weighing 60 tons, was detonated for the first time by the U.S. in the autumn of 1952. It yielded energy equivalent to 5 megatons of TNT.

It is likely that in all hydrogen bombs in the U.S., U.S.S.R., and British stockpiles use is made of a compound containing lithium. Lithium and hydrogen form the compound lithium hydride, which is a solid substance under ordinary conditions, and stable so long as it does not come into contact with water. Lithium deuteride is a similar solid substance. A hydrogen bomb can be made by putting a thousand pounds of lithium deuteride around an ordinary atomic bomb, the detonator. Such a bomb was first detonated by the U.S.S.R., in August 1953.

Also, the isotope of lithium that contains six nucleons reacts with a neutron to produce helium and tritium, and the tritium produced in this way can then react with the deuterium to produce more helium, liberating additional neutrons. Deuterium can be made in unlimited amounts by fractionating ordinary water. Lithium can be obtained in unlimited amounts from the ores of lithium, such as lepidolite. Hydrogen bombs probably can be made very cheaply.

Superbombs

Hydrogen bombs sound like the answer to the generals' prayer—they are cheap, they can be made from rather easily available materials in unlimited number, and they can be made as big as desired, big enough to take care of any target in the world.

Nevertheless, it has been found possible to do still better. Superbombs are even cheaper, and they can do anything that a hydrogen bomb can do and something else besides.

In the process of making uranium-235 or of making plutonium-239, ordinary uranium is depleted of its uranium-235. The residue, uranium-238, was for a while a drug on the market. There was no use to which it could be put.

It cannot be used to make an ordinary atomic bomb. However, it will undergo fission under the conditions of extremely high temperatures that exist when a hydrogen bomb is detonated.

It is accordingly possible to convert a hydrogen bomb into a superbomb in a simple way: by surrounding it with a shell of uranium-238 metal (or of ordinary uranium metal, uranium-238 containing a small amount of uranium-235). In this way the explosive power of the bomb may be doubled or tripled at little additional cost. And, moreover, the uranium-238, when it undergoes fission, produces great amounts of radioactive fission products, which may be effective in killing off additional millions of people in the enemy country.

The superbomb is a fission-fusion-fission bomb. When a superbomb is detonated the first process that occurs is the fission of a few pounds of uranium-235 or plutonium-239, with the liberation of energy equivalent to 20,000 tons or 100,000 tons of TNT. The temperature is raised to about fifty million degrees, and about a thousand pounds of lithium deuteride may then undergo the fusion reaction, liberating ten megatons TNT equivalent of energy. At this high temperature a thousand pounds of uranium-238 undergoes fission, liberating an additional ten megatons of explosive energy, and also producing a great amount of radioactive fission products.

Several superbombs have been detonated during recent years. It has been estimated that the first superbomb, exploded by the United States at Bikini on 1 March 1954, had a fusion stage of about five megatons and a fission stage, the third stage, of about twelve megatons. The largest Russian superbombs, including those of March 1958, seem to have been of about the same size.

These hydrogen bombs and superbombs are the weapons that will be used in a Third World War, if there ever is such a war. I cannot believe that man is so lacking in intelligence, so lacking in the power of reason, as to permit a Third World War, a great nuclear war, to take place.

Important attributes of the new weapons are their production of great amounts of high-energy radiation at the moment of explosion and also their production of radioactive materials that give rise to high-energy radiation at later times. These attributes add to the present world danger . . . .

MARY LUCY CARTWRIGHT

1900-1998

"Mathematics and Thinking Mathematically"

1969

Mary Cartwright (1900-1998) was born in Northamptonshire, England, the daughter of a rector. When she entered St Hugh's College, Oxford, she was one of only five women at that university studying mathematics. On graduation in 1923, she opted to teach mathematics for four years to help her family's finances. Then in 1928, she returned to Oxford and, under the supervision of G. H. Hardy (See *A Mathematician's Apology* in this volume) and E. C. Titchmarsh, completed her Doctorate of Philosophy in mathematics in only two years. In the early 1930s, she went to Cambridge on a research fellowship where she remained for the rest of her career as a lecturer and then a reader in mathematics.

During her career, she published over 100 mathematical papers and received many mathematical honors. In the 1940s she worked on what was to become chaos theory with John Littlewood. At that time, researchers were trying to explain mathematically why radios being used by soldiers during WW II had failed. In 1969 she was knighted, making her Dame Mary Cartwright. Marked by a dry sense of humor, the new

"Dame" Cartwright was asked by a colleague if now, since she had been knighted, they would have to bow down to her three times. Mary replied "No. Twice will do."

The essay "Mathematics and Thinking Mathematically" was originally presented at Goucher College, Baltimore, MD, in 1969. In it she explores the concepts of mathematical and abstract thinking, how these thought processes come about, and how they can be nurtured. The process by which we learn mathematics has been the subject of much study but, as elementary school teachers testify, getting young students to think abstractly is a challenge. The observations presented by Dame Cartwright may help both those seeking to teach and those developing their own thinking abilities.

Sources:

Ayoub, Raymond. June 2004. *Musings of the Masters: An Anthology of Miscellaneous Reflections*. Mathematical Association of America. 6-16.

Davis, Philip J. 1998. "Snapshots of a Lively Character: Mary Lucy Cartwright, 1900-1998." *Society for Industrial and Applied Mathematics*. Retrieved February 13, 2007, from http://www.siam.org/news/news.php?id=863.

Riddle, Larry. 2006. "Dame Mary Lucy Cartwright." *Biographies of Women Mathematicians*. Retrieved February 13, 2007, from http://www.agnesscott.edu/Lriddle/WOMEN/cartwght.htm.

Selection From:

Cartwright, Mary Lucy. 1969. "Mathematics and Thinking Mathematically." In Raymond Ayoub, Ed. *Musings of the Masters: An Anthology of Miscellaneous Reflections*. Mathematical Association of America. 2004. 6-16.

Mathematics and Thinking Mathematically

This year I find myself in the Division of Applied Mathematics at Brown University. Not every University or College has a separate department of applied mathematics, but for over thirty years I was classed as a pure mathematician in the University of Cambridge, England, and some applied mathematicians there prefer to call themselves theoretical physicists. Moreover, I once heard a geophysicist with a mainly mathematical training say that he used 'applied mathematics' as a term of abuse, meaning stuff which was not good mathematics and not really relevant to any physical problem. All these factors have made me think about the borderline between mathematics and its applications, not only to physical problems of a more or less traditional type, but also to statistical, economic, and industrial problems.

It is well known that the origins of some of the most abstract pure mathematics can be traced through the theory of Fourier series to a problem about vibrating strings, or through the theory of irrational numbers to Greek geometry and Egyptian devices for measurement of right angles, but the pure mathematicians of the last 100 or 150 years have been pursuing the mathematics for its own sake without any thought of vibrating strings. On the other hand many major new developments in pure mathematics were initiated quite specifically for the purpose of using them in some application. For instance this is certainly true of Newton's contributions to the calculus, and of probability theory, and this still seems to be happening in operations research and control theory. In distinguishing pure mathematics from applied two questions seem to arise. Is the work truly abstract and separated from all applications? And is it any more mathematical if it is truly abstract and pursued strictly for its own sake?

If we delve into the beginnings of mathematical thought in very young children or primitive peoples, there is plenty of evidence to show that the power of complete abstraction comes very slowly, and indeed to many people it probably only ever comes in a very restricted sense. A number of eminent people take the view that thought begins with the idea of actions performed in the mind only, that is to say operations. According to Piaget an ordinary child, by the time the child is two, can work out how he is going to do something before he does it, *provided* that the situation is simple and is familiar to him, but in order to understand abstract mathematical concepts such as 1, 2, 3, 4, . . . , the child has to move from perceptions arising from his environment and actions to the abstractions, a long and gradual process. Much work has been done by Piaget and Innhelder on the child's conception of space, and, for instance, its powers to distinguish between different kinds of figures such as a circle, a square, and a circle with a little one either inside or outside. Their experiments have thrown much light on the development of numerical, spatial, and physical concepts of a very elementary kind among young children, but it seems doubtful to me whether the abilities tested are always truly mathematical. For young blackbirds will gape at a piece of black cardboard consisting of one large circle and two small ones attached to it, but they only gape at the small circle whose size is a certain proportion of that of the large circle. This indicates that the ability to distinguish between certain shapes may have psychological foundations.

H. and H. A. Frankfort in an essay on myth and reality point out that ancient man could reason and work out the causes of things, but worked on very different hypotheses from ours. The primitive mind asks 'who' when it looks for a cause, and cannot withdraw far from perceptual reality. When the river does not rise, the river has *refused* to rise, and so the river or the gods intend to convey something to the people. At the same time primitive man used symbols much as we do, but he can no more conceive them as signifying, yet separate from, the gods or powers than he can consider a relationship—such as resemblance—as connecting, and yet separate from, the objects compared. Hence there is a coalescence of the symbol and what it signifies, as there is coalescence of two objects compared so that one may stand for the other.

Frankfort then gives an example of this coalescence in which pottery bowls with the names of hostile tribes were solemnly smashed at a ritual by the Egyptians in the belief that real harm was done to the enemies by the destruction of their names. It may seem a far cry from this to modern mathematics, but Bochner has drawn a parallel between mathematics and myth, and replaced myth by mathematics in some of Frankfort's sentences. I am not prepared to go as far as he does by replacing the word myth by mathematics in a sentence which then asserts that mathematics transcends reasoning in that it wants to bring about the truth it proclaims. However, in the ritual we have two fundamental features of mathematics, symbols representing something and operations on those symbols representing operations on the thing itself. Symbols and notation are part of the essential basis of mathematics, and I believe that the development and standardization of a good notation is an extremely important part of the development of mathematics.

If we turn to the extreme other end of the scale, we run into another kind of difficulty in separating the mathematics from its applications. Some pure mathematicians seem to do their mathematical thinking in terms of idealized physical and spatial ideas. The late G. H. Hardy, who taught me, was very much against applied mathematics, but in a footnote to a joint paper with J. E. Littlewood published in a Swedish periodical he wrote that a certain problem is most easily grasped in terms of cricket averages. Norbert Wiener would translate a mathematical problem into the language of Brownian motion, and I believe that his thinking was completely abstract although I do not know the theory, or remember what he said well enough to be quite sure. Hadamard has described his visualization of the proof that there is a prime greater than 11. To consider all prime numbers from 2 to 11, i.e., 2, 3, 5,

7, 11 he visualized a confused mass. Forming the product 2 x 3 x 7 x 11=N, since N is large, he visualized a point remote from the mass. Increasing the product by 1 he saw another point a little beyond the first. $N + 1$, if not a prime, is divisible by a prime greater than 11; Hadamard saw a place between the mass and the first point. This seems to me to be a sort of mathematical shorthand and would certainly have to be translated back to numbers before it could be communicated to anyone else.

As I said earlier, I have until now always been classed as a pure mathematician, but Professor J. E. Littlewood and I did a lot of work on the theory of ordinary differential equations arising from problems of radio engineering. Littlewood is also a very pure mathematician in many ways, but he worked on anti-aircraft gunfire in the First World War, and he translated our problems, which were suggested by radio values and oscillations, capacitance and inductance, etc., into dynamical problems and called all the solutions of our equations 'trajectories' as if they were the paths of missiles shot from a gun. In the radio problems there are oscillations with negative damping, and so we had periodic trajectories going up and down over and over again, and I am sure that the abstraction was complete although there was often a certain woolliness until the argument was complete, just as in Hadamard's visualization. Between these two extremes there are sonic users of complicated mathematics, physicists and engineers in particular, who are thinking all, or nearly all, the time in terms of the physics of the problem. Engineers have consulted me about a number of different types of problem, radio, control theory, oscillations of stretched wires; they usually come with some equations and very little explanation. I have to ask a lot of questions before they tell me everything relevant to the mathematical problem. It seems difficult for them to think in abstract mathematical terms, the symbols to them seem to mean the engineering concepts, currents and circuit constants such as impedance and inductance. This is important in two ways. The engineers have mental reservations and can check at every stage because they visualize how the physical system works. On the other hand they find it difficult to apply the mathematical processes used in one field to any other physical problem, even if they are just as relevant there. Some years ago at a conference for engineers I was asked to speak on Liapunov's method for stability problems. I described the basic principles as simply as I could, and after I spoke Professor Parks lectured on applications of the method. Many in the audience commented that the order of our lectures should have been reversed, and that they would have understood my lecture much better if they had understood that I was talking about the phase plane. It is possible that this was partly a question of notation and terminology, but I believe that they could do advanced mathematics best by thinking of it in terms of their particular engineering problems. The Liapunov method was developed mainly in connection with control engineering and by now has adopted much of its terminology, but the mathematics arising there need to be abstracted and put in a form which makes it available in connection with other applications. Problems of ordinary differential equations have arisen in connection with astronomy, ballistics, radio engineering, control theory, mechanical oscillations of machinery; each application has special features, and the theory of it was often developed in a correct logical form quite a long way before it was fitted into the general theory of ordinary differential equations as pure mathematics. The individual who formulated the equation and asked the question is, in the sense of my title, thinking mathematically, but he is not doing mathematics until he operates on his symbols. Please note that I do not say 'asked for a solution of the equation' because, although he may say that, he really wants to know something about the solutions in general. Is there a periodic solution? Is it stable? Will it remain stable if I change a certain parameter? Will the period be longer or shorter? He may find the methods which he needs in the literature and do the work himself. He may find a mathematician to help him. Although I myself have helped to develop the general theory and settle certain theoretical problems, I do not think that I have ever produced a result useful for any specific practical problem when it was needed. For soon after Littlewood and I began work on these problems, it was realized that the variations in individual thermionic valves was so great that precise mathematical results were not worth the trouble, and satisfactory experimental determinations could be more easily obtained. In recent times the person who formulates the mathematical statement of a physical or other real life problem usually does not do anything very original in the mathematical handling of it, although some interesting purely mathematical work on matrices appears in journals concerned with computing or applications to economics, detached from other pure mathematics.

To sum up so far I believe that the dividing line between strictly abstract thinking in mathematics and thinking in terms of the real world is by no means clearly defined and some of the major developments in mathematics such as the calculus were thought out more or less in terms of the real world. Further abstraction does not necessarily make the mathematics any better. For the Babylonian schoolmasters constructed sets of most complicated artificial formulae, perhaps 200 on one tablet, for their pupils to simplify. Their mathematics was sufficiently abstract for them to be indifferent whether they added the number of men to the number of days. In present circumstances this seems abstraction at its worst, but perhaps then it was a step forward. The Babylonians must have developed the laws of arithmetic a long way to set these complicated exercises, but mainly for practical purposes whether it was accounting or astronomy.

Now let us turn to those who do mathematics for its own sake. I should like to begin with the Hindu who in about 1200 B.C. wrote, "As crests on the heads of peacocks, as the gems on the hoods of snakes, so is ganita, mathematics, at the top of the sciences known as the Vedanga." Ganita is literally the science of calculation and in the early days it consisted of finger arithmetic, mental arithmetic, and higher arithmetic in general. At first it included astronomy, but geometry belonged elsewhere. At one stage higher mathematics was called 'dust work' because it was done in sand spread on the board or on the ground. We owe our so-called Arabic numerals to the Hindus, and they advanced a long way in algebra very early.

Most people consider that the Greeks were the first to do mathematics for its own sake and to realize the need for proof. The word 'mathema' meant originally a subject of instruction, but very early it was restricted to mathematical subjects among which Pythagoras included geometry, theory of numbers, sphaeric (or spherical trigonometry used for astronomy), and music. They classified numbers not only as odd and even, but as even-even, 2^m; even-odd, $2(2n+1)$; odd-even $2^{m+1}(2n+1)$, and also proved that there are an infinity of primes. I doubt whether they could calculate as well as the Babylonians, but probably that did not attract them, and also they lacked the incentives provided by the government of a far flung empire. I feel that I have to remind myself of the difficulties due to the absence of convenient symbols. Sir Thomas Heath writing of the arithmetic of Nicomachus said 'If the verbiage is eliminated, the mathematical content can be stated in quite a small compass,' but Heath used modern notation and Arabic numerals. In the *Wasps* of Aristophanes one of the characters tells his father to do an easy sum 'not with pebbles but with fingers,' and Herodotus says that, in reckoning with pebbles, Greeks move left to right, Egyptians right to left, which implies vertical columns facing the reckoner.

The Greeks also developed a theory of geometry which remained more important than any other for nearly 2,000 years, and was the first deliberate development of a logical system in mathematics. In the third century A.D. an unknown writer jokingly used words of Homer intended for something else to describe mathematics:

> Small at her birth, but rising every hour.
> She stalks on earth and shakes the world around.

For, says Anatolius, Bishop of Laodacia, who quoted it, mathematics begins with a point and a line and forthwith it takes in the heaven itself and all things within its compass. If this was the Greek viewpoint at such a late date, is it possible that their geometry was not truly abstract and that the symbols of point and line were still partly coalesced with the abstract point and line?

The position of geometry and more generally spatial concepts in mathematics is not completely clear to me. In recent times all types of geometry have been given an analytical basis and freed from the logical difficulties such as those which used to worry schoolmasters teaching about congruent triangles by the method of superposition. I therefore ask myself whether geometry and spatial concepts are really part of the basis of mathematics or a field of application similar to mechanics, both terrestrial and celestial, or to games of chance. The reason for the traditional special position of geometry may be that in geometry the symbols are the objects themselves; the abstract point, line, and triangle are represented by a point, line, and triangle; what is more, so long as the geometry is plane geometry they can be drawn on a flat surface by pen or pencil on paper or in sand on the ground. When Greek geometry was being developed there was no good notation for dealing with numbers, and even in the 15th Century the solution of a cubic equation was described in geometrical terms and illustrated by a figure for lack of a good algebraic notation. In mechanics a comparable real life representation of motion could not be used to explain the theory; written symbols or geometrical figures were needed for communication. But if we ask whether the contributions of spatial concepts to modern mathematics are greater than those of other real life problems it is difficult to answer. Spatial thinking has led to the highly abstract theory of irrational numbers of Cantor and Dedekind, and permeates mathematical thought in almost all fields; the physical sciences have given rise to the calculus (not without the help of geometry), and statistics and probability have their basis in multitudinous practical problems.

Pfeiffer explains the situation well in relation to probability. Some of the salient points in his account are as follows: The history of probability theory (as is true of most theories) is marked both by brilliant intuition and discovery and by confusion and controversy. Until certain patterns had emerged to form the basis of a clear-cut theoretical model, investigators could not formulate problems with precision, and reason about them with mathematical assurance.

From what some people say it sounds to me as if quantum theory had not yet reached this stage, but it is certainly beyond my competence to form a valid judgment.

Pfeiffer continues by saying that although long experience was needed to produce a satisfactory theory, we need not retrace and relive the fumblings which delayed the discovery of an appropriate mathematical model. That is, a mathematical system whose concepts and relationships correspond to the appropriate concepts and relationships of the real world. Once the model has been discovered, studied, and refined, it becomes possible for an ordinary mind to grasp, in a reasonably short time, a pattern which took decades of effort and the insight of genius to develop in the first place. I note that Pfeiffer asserts that the most successful model of probability theory known at present is characterized by considerable mathematical abstractness.

J. Willard Gibbs wrote 'One of the principal objects of theoretical research in any department of knowledge is to find the point of view from which the subject appears in its greatest simplicity' and Bushaw says that one of the distinctive characteristics of modern mathematics is its way of taking old mathematical ideas apart like watches, studying the parts separately, and putting these parts together again in new and interesting combinations and studying these complications in turn. I believe that this process has contributed enormously to this simplification in mathematics itself, and so made it more readily available for applications. Mandelbrojt referring to the quotation from Willard Gibbs says 'Integration in function spaces provided such a point of view over and over again in widely scattered areas of knowledge and it gave us not only a new way of looking at problems but actually a new way of thinking about them.' Now one might call Fréchet the father of abstract spaces, and in the front of his book he puts a quotation from Hadamard's survey of functional analysis given in 1911. 'The functional continuum does not present any simple concept to our imagination. Geometrical intuition tells us nothing *a priori* about it. We are forced to remedy this ignorance and we can do it only analytically, by creating a chapter of the theory of sets for handling the functional continuum.' Elsewhere Hadamard wrote that the calculus of variations was nothing but the first chapter of functional analysis, and of his own work on the calculus of variations, hyperbolic partial differential equations, and certain other topics he said that he owed the greater part to his contacts with the physicist Duhem, through Duhem's book on hydrodynamics, elasticity, and acoustics and many conversations when they were both at Bordeaux. So we have a record here of the complete cycle from a physical basis through the calculus of variations to functional analysis and abstract spaces, and thence to a multitude of applications through the process of analyzing geometrical ideas and putting them together again in a most abstract new way to create function spaces.

A further variation on this pattern has become evident of recent years and that is the use of an auxiliary model consisting of various graphical, mechanical, and other aids to visualizing, remembering, and even discovering things about the mathematical model. The visual images of Hadamard, Hardy's cricket averages, and Littlewood's trajectories might be considered as auxiliary models, but of more universal significance are the analogue machines with electronic devices which simulate what happens in, for instance, fluid mechanics, or rather what corresponds in the mathematical model. We now have

(A) The real world of actual phenomena, known to us by various ways of experiencing these phenomena.
(B) The abstract world of the mathematical model which uses symbols to state relationships and facts with great precision and economy.
(C) The auxiliary model.

The transition from A to B is the formulation of real world phenomena in mathematical terms; the transition B to A is the interpretation of the deduction by pure mathematics from that formulation. Both these I consider to be thinking mathematically, but only the deductions inside B are mathematics. We may also think mathematically by moving from B to C which is a secondary interpretation, and then either back to B to confirm what C has suggested or from C direct to A.

As Pfeiffer points out, the value of both the mathematical model and the auxiliary model depends on how successfully the appropriate features of the model may be related to the 'real-life' situation. The models cannot be used to prove anything about the real world, although a study of it may help us to discover important facts about the real world. A model is not true or false; it fits or it does not fit. It is unsatisfactory if either (1) the solutions of the model problems have unrealistic interpretations, for instance, arbitrarily large quantities or arbitrarily fine differences, or (2) it is incomplete or inconsistent so that the mathematics produces contradictions. Many models fit amazingly well. Karl Pearson wrote 'The mathematician, carried along on his flood of symbols, dealing apparently with purely formal truths, may still reach results of endless importance for our description of the physical universe.

Until perhaps 100 years ago many scientists and mathematicians knew a bit of everything, and the mathematical formulation, as I said of Newton in particular, was done by someone who was a good enough mathematician to develop the mathematics to a considerable extent. This is particularly true of Sir Isaac Newton, but in these days of specialization the scientist or economist, or worker in close contact with the real world situation must do stage A→B. Sir Cyril Hinshelwood, former President of the Royal Society, said 'Scientists need to be taught mathematics as a language they can actually speak. It is of great importance for the scientist to be able to learn the art of formulating problems in mathematical terms which of course is a quite difficult job. You have to think very accurately and carefully about a problem before you can do it. You have to have practice in speaking the language of mathematics. It does not matter being an expert in differential equations. You can go to the expert for help in solving an equation. But you cannot expect the mathematician to do the translation into mathematics. There should be an early and rather intensive cultivation of the power of thinking about real things and the application of mathematical symbolism to physical ideas.' He went on to draw a parallel between learning simple French as a child and learning to express physical ideas in mathematics when the level of physics and mathematics reached are both elementary, so that the child becomes accustomed to the process by easy stages. Although he advocates, as I do, that the scientist should do the mathematical formulation, his words seem to imply an incomplete abstraction. In his mind the mathematical symbols were still representing their physical counterparts, not that this matters for a scientist who has access to an expert mathematician, but it is clear from Mandelbrojt's remarks on function spaces and Hadamard's remarks about the functional continuum that without complete abstraction on the part of some mathematicians we should lack some of the most expressive parts of the mathematical language used by scientists.

'Euclid's geometry was supposed to deal with real objects, whether in the physical world or in some ideal world. The definitions which preface several books in the *Elements* are supposed to communicate what object the author is talking about even though, like the famous definition of the point and the line, they may not be required in the sequel. The fundamental importance of the advent of non-Euclidean geometry is that by contradicting the axiom of parallels it denied the uniqueness of geometrical concepts and hence, their reality. By the end of the nineteenth century, the interpretation of the basic concepts of geometry had become irrelevant. This was the more important since geometry had been regarded for a long time as the ultimate foundation of all mathematics. However, it is likely that the independent development of the foundations of the number system which was sparked by the intricacies of analysis would have deprived geometry of its predominant position anyhow.'

Although it confirms my views on Euclidean geometry, it does not seem to recognize the geometrical origin of the theory of irrational numbers.

I also noticed that A. Aaboe in *Episodes from the Early History of Mathematics,* writes 'Even the oft repeated statement that the Egyptians knew the 3, 4, 5 right angle has no basis in available texts, but was invented about 80 years ago.'

References

1. S. Bochner, *The Role of Mathematics in the Rise of Science,* Princeton, 1966.
2. D. Bushaw, *Elements of General Topology,* Wiley, New York, 1963.
3. B. Datta and A. N. Singh, *A History of Hindu Mathematics,* Lahore, 1935.
4. H. and H. A. Frankfort, *The Intellectual Adventures of Ancient Man,* Chicago, 1946.
5. T. L. Heath, *A History of Greek Mathematics,* Vol. 2, Oxford, 1921.
6. S. Mandelbrojt, Les Tauberiens Généraux de Norbert Wiener, *Bull. Amer. Math. Soc.,* 72 (1966) 48-51.

7. O. Neugebauer, The Exact Sciences in Antiquity, *Acta Hist. Sci. Nat. Medicinalium*, Copenhagen, 9 (1951).

8. P. E. Pfeiffer, *Concepts of Probability Theory*, McGraw-Hill, New York, 1965.

9. J. Piaget, B. Inhelder, and A. Sjeminska, *A Child's Conception of Geometry*, Trans by E. A. Lunzer, Basic Books, New York, 1960.

10. N. Tinbergen, *The Herring Gull's World*, Basic Books, New York, 1961.

DOUGLAS R. HOFSTADTER

1945-

Gödel, Escher, Bach: an Eternal Golden Braid

1979

The breadth of Douglas R. Hofstadter's knowledge is reflected in his current position at Indiana University: College Professor of Cognitive Science and Computer Science, and Adjunct Professor of History and Philosophy of Science, Philosophy, Comparative Literature, and Psychology. He heads the Fluid Analogies Research Group (FARG) at Indiana University's Center for Research on Concepts and Cognition (CRCC). Its work involves the attempt to build computer models of human thinking, including "perception, analogical thought, and creativity. The FARG members engage in creative intellectual endeavors, either scientific or artistic, such as poetry translation, discovery in mathematics, the study of human error making, the study of humor, the study of sexist language and imagery, the creation of various types of art, and so on" (see Hofstadter site).

He is best known for his first work, *Gödel, Escher, Bach: an Eternal Golden Braid*, (more often known as "GEB"), published in 1979, four years after his doctorate was awarded by Oregon State. It won a Pulitzer Prize in 1980. At the top of the New York Times best seller list, the book has become a "must read" for students of artificial intelligence. GEB is, on one level, about how the creative achievements of logician Kurt Gödel, artist M. C. Escher, and composer Johann Sebastian Bach interweave; on a more abstract level, the book addresses the question of consciousness, the possibility of artificial intelligence, and the attempts to discover what "self" really means.

In the first chapter, "The MU-puzzle" (included here), Hofstadter presents the reader with a puzzle that is easy to understand but not so easy to solve.

Source:

Hofstadter, Douglas. 2006. Indiana University Cognitive Science Program. Retrieved February 21, 2007, from http://www. cogs.indiana.edu/people/homepages/hofstadter.html.

Worthy, Glen. 2006. "Douglas R. Hofstadter: Analogy Core, Core as Analogy." Stanford University Libraries. Retrieved February 21, 2007, from http://prelectur.stanford.edu/ lecturers/hofstadter/.

Selection From:

Hofstadter, Douglas. 1980. *Gödel, Escher, Bach: an Eternal Golden Braid.* New York: Vintage. 33-41.

CHAPTER 1

The MU-puzzle

FORMAL SYSTEMS

ONE OF THE most central notions in this book is that of a *formal system*. The type of formal system I use was invented by the American logician Emil Post in the 1920's, and is often called a "Post production system". This Chapter introduces you to a formal system and moreover, it is my hope that you will want to explore this formal system at least a little; so to provoke your curiosity, I have posed a little puzzle.

"Can you produce MU?" is the puzzle. To begin with, you will be supplied with a *string* (which means a string of letters).* Not to keep you in suspense, that string will be MI. Then you will be told some rules, with which you can change one string into another. If one of those rules is applicable at some point, and you want to use it, you may, but—there is nothing that will dictate which rule you should use, in case there are several applicable rules. That is left up to you—and of course, that is where playing the game of any formal system can become something of an art. The major point, which almost doesn't need stating, is that you must not do anything which is outside the rules. We might call this restriction the "Requirement of Formality". In the present Chapter, it probably won't need to be stressed at all. Strange though it may sound, though, I predict that when you play around with some of the formal systems of Chapters to come, you will find yourself violating the Requirement of Formality over and over again, unless you have worked with formal systems before.

The first thing to say about our formal system—the *MIU-system*—is that it utilizes only three letters of the alphabet: M, I, U. That means that the only strings of the MIU-system are strings which are composed of those three letters. Below are some strings of the MIU-system:

---

* In this book, we shall employ the following conventions when we refer to strings. When the string is in the same typeface as the text, then it will be enclosed in single or double quotes. Punctuation which belongs to the sentence and not to the string under discussion will go outside of the quotes, as logic dictates. For example, the first letter of this sentence is 'F', while the first letter of 'this sentence' is 't'. When the string is in **Quadrata Roman**, however, quotes will usually be left off, unless clarity demands them. For example, the first letter of **Quadrata** is Q.

> MU
> UIM
> MUUMUU
> UIIUMIUUIMUIIUMIUUIMUIIU

But although all of these are legitimate strings, they are not strings which are "in your possession". In fact, the only string in your possession so far is MI. Only by using the rules, about to be introduced, can you enlarge your private collection. Here is the first rule:

RULE I: If you possess a string whose last letter is I, you can add on a U at the end.

By the way, if up to this point you had not guessed it, a fact about the meaning of "string" is that the letters are in a fixed order. For example, MI and IM are two different strings. A string of symbols is not just a "bag" of symbols, in which the order doesn't make any difference.

Here is the second rule:

RULE II: Suppose you have Mx. Then you may add Mxx to your collection.

What I mean by this is shown below, in a few examples.

> From MIU you may get MIUIU.
> From MUM, you may get MUMUM.
> From MU, you may get MUU.

So the letter 'x' in the rule simply stands for any string; but once you have decided which string it stands for, you have to stick with your choice (until you use the rule again, at which point you may make a new choice). Notice the third example above. It shows how, once you possess MU, you can add another string to your collection: but you have to get MU first! I want to add one last comment about the letter '*x*': it is not part of the formal system in the same way as the three letters 'M', 'I', and 'U' are. It is useful for us, though, to have some way to talk in general about strings of the system, symbolically—and that is the function of the '*x*': to stand for an arbitrary string. If you ever add a string containing an '*x*' to your "collection", you have done something wrong, because strings of the MIU-system never contain "*x*"s!

Here is the third rule:

RULE III: If III occurs in one of the strings in your collection, you may make a new string with U in place of III.

Examples:
        From UMIIIMU, you could make UMUMU.
        From MIIII you could make MIU (also MUI).
        From IIMII, you can't get anywhere using this rule.
                (The three I's have to be consecutive.)
        From MIII, make MU.

Don't, under any circumstances, think you can run this rule backwards, as in the following example:

        From MU, make MIII. <= This is wrong.

Rules are one-way.

        Here is the final rule:

RULE IV: If UU occurs inside one of your strings, you can drop it.

        From UUU, get U.
        From MUUUIII, get MUIII.

        There you have it. Now you may begin trying to make MU. Don't worry if you don't get it. Just try it out a bit—the main thing is for you to get the flavor of this MU-puzzle. Have fun.

Theorems, Axioms, Rules

The answer to the MU-puzzle appears later in the book. For now, what is important is not finding the answer, but looking for it. You probably have made some attempts to produce MU. In so doing, you have built up your own private collection of strings. Such strings, producible by the rules, are called *theorems*. The term "theorem" has, of course, a common usage in mathematics which is quite different from this one. It means some statement in ordinary language which has been proven to be true by a rigorous argument, such as Zeno's Theorem about the "unexistence" of motion, or Euclid's Theorem about the infinitude of primes. But in formal systems, theorems need not be thought of as statements—they are merely strings of symbols. And instead of being *proven,* theorems are merely *produced,* as if by machine, according to certain typographical rules. To emphasize this important distinction in meanings for the word "theorem", I will adopt the following convention in this book: when "theorem" is capitalized, its meaning will be the everyday one—a Theorem is a statement in ordinary language which somebody once proved to be true by some sort of logical argument. When uncapitalized, "theorem" will have its technical meaning: a string producible in some formal system. In these terms, the MU-puzzle asks whether MU is a theorem of the MIU-system.

I gave you a theorem for free at the beginning, namely MI. Such a "free" theorem is called an *axiom*—the technical meaning again being quite different from the usual meaning. A formal system may have zero, one, several, or even infinitely many axioms. Examples of all these types will appear in the book.

Every formal system has symbol-shunting rules, such as the four rules of the MIU-system. These rules are called either *rules of production* or *rules of inference.* I will use both terms.

The last term which I wish to introduce at this point is *derivation.* Shown below is a derivation of the theorem MUIIU:

(1) MI   axiom
(2) MII   from (1) by rule II
(3) MIIII   from (2) by rule II
(4) MIIIIU   from (3) by rule I
(5) MUIU   from (4) by rule III
(6) MUIUUIU   from (5) by rule II
(7) MUIIU   from (6) by rule IV

A derivation of a theorem is an explicit, line-by-line demonstration of how to produce that theorem according to the rules of the formal system. The concept of derivation is modeled on that of proof, but a derivation is an austere cousin of a proof. It would sound strange to say that you had *proven* MUIIU, but it does not sound so strange to say you have *derived* MUIIU.

Inside and Outside the System

Most people go about the MU-puzzle by deriving a number of theorems, quite at random, just to see what kind of thing turns up. Pretty soon, they begin to notice some properties of the theorems they have made; that is where human intelligence enters the picture. For instance, it was probably not obvious to you that all theorems would begin with M, until you had tried a few. Then, the pattern emerged, and not only could you see the pattern, but you could understand it by looking at the rules, which have the property that they make each new theorem inherit its first letter from an earlier theorem; ultimately, then, all theorems' first letters can be traced back to the first letter of the sole axiom MI—and that is a proof that theorems of the MIU-system must all begin with M.

There is something very significant about what has happened here. It shows one difference between people and machines. It would certainly be possible—in fact it would be very easy—to program a computer to generate theorem after theorem of the MIU-system; and we could include in the program a command to stop only upon generating U. You now know that a computer so programmed would never stop. And this does not amaze you. But what if you asked a friend to try to generate U? It would not surprise you if he came back after a while, complaining that he can't get rid of the initial M, and therefore it is a wild goose chase. Even if a person is not very bright, he still cannot help making some observations about what he is doing, and these observations give him good insight into the task—insight which the computer program, as we have described it, lacks.

Now let me be very explicit about what I meant by saying this shows a difference between people and machines. I meant that it is possible to program a machine to do a routine task in such a way that the machine will never notice even the most obvious facts about what it is doing; but it is inherent in human consciousness to notice some facts about the things one is doing. But you knew this all along. If you punch "1" into an adding machine, and then add 1 to it again, and then add 1 again, and again, and again, and continue doing so for hours and hours, the machine will never learn to anticipate you, and do it itself, although any person would pick up the repetitive behavior very quickly. Or, to take a silly example, a car will never pick up the idea, no matter how much or how well it is driven, that it is supposed to avoid other cars and obstacles on the road; and it will never learn even the most frequently traveled routes of its owner.

The difference, then, is that it is *possible* for a machine to act unobservant; it is impossible for a human to act unobservant. Notice I am not saying that all machines are necessarily incapable of making sophisticated observations; just that some machines are. Nor am I saying that all people are always making sophisticated observations; people, in fact, are often very unobservant. But machines can be made to be totally unobservant; and people cannot. And in fact, most machines made so far are pretty close to being totally unobservant. Probably for this reason, the property of being unobservant seems to be the characteristic feature of machines, to most people. For example, if somebody says that some task is "mechanical", it does not mean that people are incapable of doing the task; it implies, though, that only a machine could do it over and over without ever complaining, or feeling bored.

Jumping out of the System

It is an inherent property of intelligence that it can jump out of the task which it is performing, and survey what it has done; it is always looking for, and often finding, patterns. Now I said that an intelligence can jump out of its task, but that does not mean that it always will. However, a little prompting will often suffice. For example, a human being who is reading a book may grow sleepy. Instead of continuing to read until the book is finished, he is just as likely to put the book aside and turn off the light. He has stepped "out of the system" and yet it seems the most natural thing in the world to us. Or, suppose person A is watching television when person B comes in the room, and shows evident displeasure with the situation. Person A may think he understands the problem, and try to remedy it by exiting the present system (that television program), and flipping the channel knob, looking for a better show. Person B may have a more radical concept of what it is to "exit the system"—namely to turn the television off! Of course, there are cases where only a rare individual will have the vision to perceive a system which governs many peoples' lives, a system which had never before even been recognized as a system; then such people often devote their lives to convincing other people that the system really is there, and that it ought to be exited from!

How well have computers been taught to jump out of the system? I will cite one example which surprised some observers. In a computer chess tournament not long ago in Canada, one program—the weakest of all the competing ones—had the unusual feature of quitting long before the game was over. It was not a very good chess player, but it at least had the redeeming quality of being able to spot a hopeless position, and to resign then and there, instead of waiting for the other program to go through the boring ritual of checkmating. Although it lost every game it played, it did it in style. A lot of local chess experts were impressed. Thus, if you define "the system" as "making moves in a chess game", it is clear that this program had a sophisticated, preprogrammed ability to exit from the system. On the other hand, if you think of "the system" as being "whatever the computer had been programmed to do", then there is no doubt that the computer had no ability whatsoever to exit from that system.

It is very important when studying formal systems to distinguish working *within* the system from making statements or observations *about* the system. I assume that you began the MU-puzzle, as do most people, by working within the system; and that you then gradually started getting anxious, and this anxiety finally built up to the point where without any need for further consideration, you exited from the system, trying to take stock of what you had produced, and wondering why it was that you had not succeeded in producing MU. Perhaps you found a reason why you could not produce MU; that is thinking *about* the system. Perhaps you produced MIU somewhere along the way; that is working *within* the system. Now I do not want to make it sound as if the two modes are entirely incompatible; I am sure that every human being is capable to some extent of working inside a system and simultaneously thinking about what he is doing. Actually, in human affairs, it is often next to impossible to break things neatly up into "inside the system" and "outside the system"; life is composed of so many interlocking and interwoven and often inconsistent "systems" that it may seem simplistic to think of things in those terms. But it is often important to formulate simple ideas very clearly so that one can use them as models in thinking about more complex ideas. And that is why I am showing you formal systems; and it is about time we went back to discussing the MIU-system.

M-Mode, I-Mode, U-Mode

The MU-puzzle was stated in such a way that it encouraged some amount of exploration within the MIU-system—deriving theorems. But it was also stated in a way so as not to imply that staying inside the system would necessarily yield fruit. Therefore it encouraged some oscillation between the two modes of work. One way to separate these two modes would be to have two sheets of paper; on one sheet, you work "in your capacity as a machine", thus filling it with nothing but M's, I's, and U's; on the second sheet, you work "in your capacity as a thinking being", and are allowed to do whatever your intelligence suggests—which might involve using English, sketching ideas, working backwards, using shorthand (such as the letter '*x*'), compressing several steps into one, modifying the rules of the system to see what that gives, or whatever else you might dream up. One thing you might do is notice that the numbers 3 and 2 play an important role, since I's are gotten rid of in three's, and U's in two's—and doubling of length (except for the M) is allowed by rule II. So the second sheet might also have some figuring on it. We will occasionally refer back to these two modes of dealing with a formal system, and we will call them the *Mechanical mode (M-mode)* and the *Intelligent mode (I-mode)*. To round out our modes, with one for each letter of the MIU-system, I will also mention a final mode—the *Un-mode (U-mode)*, which is the Zen way of approaching things. More about this in a few Chapters.

Decision Procedures

An observation about this puzzle is that it involves rules of two opposing tendencies—the *lengthening rules* and the *shortening rules*. Two rules (1 and II) allow you to increase the size of strings (but only in very rigid, prescribed ways, of course); and two others allow you to shrink strings somewhat (again in very rigid ways). There seems to be an endless variety to the order in which these different types of rules might be applied, and this gives hope that one way or another, MU could be produced. It might involve lengthening the string to some gigantic size, and then extracting piece after piece until only two symbols are left; or, worse yet, it might involve successive stages of lengthening and then shortening and then lengthening and then shortening, and so on. But there is no guarantee of it. As a matter of fact, we already observed that U cannot be produced at all, and it will make no difference if you lengthen and shorten till kingdom come.

Still, the case of U and the case of MU seem quite different. It is by a very superficial feature of U that we recognize the impossibility of producing it: it doesn't begin with an M (whereas all theorems must). It is very convenient to have such a simple way to detect nontheorems. However, who says that that test will detect *all* nontheorems? There may be lots of strings which begin with M but are not producible. Maybe MU is one of them. That would mean that the "first-letter test" is of limited usefulness, able only to detect a portion of the nontheorems, but missing others. But there remains the possibility of some more elaborate test which discriminates perfectly between those strings which can be produced by the rules, and those which cannot. Here we have to face the question, "What do we mean by a test?" It may not be obvious why that question makes sense, or is important, in this context. But I will give an example of a "test" which somehow seems to violate the spirit of the word.

Imagine a genie who has all the time in the world, and who enjoys using it to produce theorems of the MIU-system, in a rather methodical way. Here, for instance, is a possible way the genie might go about it:

Step 1: Apply every applicable rule to the axiom Ml. This yields two new theorems: MIU,

Step 2: Apply every applicable rule to the theorems produced in step 1. This yields three new theorems: MIIU, MUM, MIIII.

Step 3: Apply every applicable rule to the theorems produced in step 2. This yields five new theorems: MIIIIU, MIIUIIU, MIUIUIUIU, MIIIIIIII, MUI.

.

. .

.

This method produces every single theorem sooner or later, because the rules are applied in every conceivable order. (See Fig. 11.) All of the lengthening-shortening alternations which we mentioned above eventually get carried out. However, it is not clear how long to wait for a given string to appear on this list,

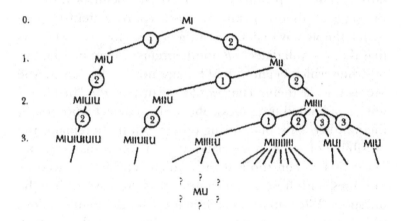

FIGURE I1. A systematically constructed "tree" of all the theorems of the MIU system. The Nth level down contains those theorems whose derivations contain exactly N steps. The encircled numbers tell which role was employed. Is MU anywhere in this tree?

since theorems are listed according to the shortness of their derivations. This is not a very useful order, if you are interested in a specific string (such as MU), and you don't even know if it has any derivation, much less how long that derivation might be.

Now we state the proposed "theoremhood-test":

> Wait until the string in question is produced; when
> that happens, you know it is a theorem—and if it
> never happens, you know that it is not a theorem.

This seems ridiculous, because it presupposes that we don't mind waiting around literally an infinite length of time for our answer. This gets to the crux of the matter of what should count as a "test". Of prime importance is a guarantee that we will get our answer in a finite length of time. If there is a test for theoremhood, a test which does always terminate in a finite amount of time, then that test is called a *decision procedure* for the given formal system.

When you have a decision procedure, then you have a very concrete characterization of the nature of all theorems in the system. Offhand, it might seem that the rules and axioms of the formal system provide no less complete a characterization of the theorems of the system than a decision procedure would. The tricky word here is "characterization". Certainly the rules of inference and the axioms of the MIU-system do characterize, implicitly, those strings that are theorems. Even more implicitly, they characterize those strings that are not theorems. But implicit characterization is not enough, for many purposes. If someone claims to have a characterization of all theorems, but it takes him infinitely long to deduce that some particular string is not a theorem, you would probably tend to say that there is something lacking in that characterization—it is not quite concrete enough. And that is why discovering that a decision procedure exists is a very important step. What the discovery means, in effect, is that you can perform a test for theoremhood of a string, and that, even if the test is complicated, it is guaranteed to terminate. In principle, the test is just as easy, just as mechanical, just as finite, just as full of certitude, as checking whether the first letter of the string is M. A decision procedure is a "litmus test" for theoremhood!

Incidentally, one requirement on formal systems is that the set of axioms must be characterized by a decision procedure—there must be a litmus test for axiomhood. This ensures that there is no problem in getting off the ground at the beginning, at least. That is the difference between the set of axioms and the set of theorems: the former always has a decision procedure, but the latter may not.

I am sure you will agree that when you looked at the MIU-system for the first time, you had to face this problem exactly. The lone axiom was known, the rules of inference were simple, so the theorems had been implicitly characterized—and yet it was still quite unclear what the consequences of that characterization were. In particular, it was still totally unclear whether MU is, or is not, a theorem.

SIV CEDERING

1939-

Letters from the Floating World

1984

A Swedish born artist now living in New York, Siv (rhymes with Steve) Cedering is a true renaissance artist producing work as an author, poet, sculptor, painter and composer. Her *Letters from the Floating World*, published in 1984, includes poems that seem to be a collection of "Letters from the Astronomers." The poems describe the political and social pressures that threatened or confined scientists in their times. Copernicus, Kepler and Galileo faced deadly opposition from the Catholic Church. Caroline Herschel, an 18th-century German, lived in a time of deep, tradition misogyny. Her "letter" reflects her disappointment that her contribution, as that of other female astronomers who have preceded her, will be erased from historic record, claimed by men with whom she worked. In more modern times, Cedering's Einstein works on his theories amid governmental folly, overkill, and ignorance about nuclear war.

Sources:

Cedering, Siv. *Siv Cedering*. Retrieved February 11, 2007, from http://www.cedering.com/.

Riddle, Larry. "Caroline Herschel." *Biographies of Women Mathematicians*. Retrieved February 11, 2007, from http://www.agnesscott.edu/Lriddle/women/herschel.htm.

Selections From:

Cedering, Siv. 1984. "Letters from the Astronomers." in *Letters from the Floating World: Selected and New Poems*. Pittsburgh: University of Pittsburgh Press. pp. 112-119.

LETTERS FROM THE ASTRONOMERS

If then the Astronomers, whereas they spie
A new-found Starre, their Opticks magnifie,
How brave are those, who with their Engine can
Bring man to heaven, and heaven again to man?
—John Donne

I. Nicholas Copernicus (1473-1543)

The sun is the center of the universe.
The planets move around the sun.
Yesterday when I went riding,
it began to snow. The seasons
change. The earth
turns
on its axis.

Share these ideas with your students,
but don't give them my name.
New continents are being discovered.
Books are being printed. Witches
burn. I am afraid
of the spirit
of the times.
What will they do, if I say
the sun is the center of this cosmic
temple, if I say
its distance from the earth
is infinitesimally small
compared to the distance
between the earth
and the stars?

I carry a pouch
of powdered unicorn's horn
and red sandal wood,
to cure
the ill. I have devised a new
monetary system.

I was asked to Rome to help create
a new calendar, but I declined.
My mathematics are not
sufficient. If you
have been trained in the art,
see if you can find the laws
that would prove
my ideas. Do you have a quadrant?
An armillary sphere? An astrolabe?

II. Johannes Kepler (1571-1630)

If we substitute the word "force"
for the word "soul," we shall have
the basic principle which lies at
the heart of my celestial physics.
                                    —Kepler

They say my mother is a witch.
She was arrested in the rectory.
They dragged her off to prison in a trunk.
They want to put her on the rack.
For weeks she has been chained.
I am writing letters
asking them to release her.
My school has been closed.
The Protestant teachers have been burned
at the stake. My youngest child
died, of smallpox.
But I try to continue my exploration
of a celestial science.
I have derived a musical scale
for each planet, from variations
in their daily motions around the sun.
A five-note scale for Jupiter.
Fourteen notes for Mercury, and Venus,
repeating her one long note.
Such harmony. As I picture each planet
floating within the geometric perfections
of space, I think geometry was implanted in man
along with the image of God.
Geometry is indeed God.

III. Galileo Galilei (1564-1642)

They say I was arrogant.
Well, I was.
I had come far
from my father's numbers
and songs.
Before me, no one had seen
the moons of Jupiter,
the phases of Venus,
the mountains of the moon.

Now I am tired of arguing with
Jesuits and Aristotelians.
I have renounced my views
on the Copernican universe.
I have denounced my students
and friends. My book
is condemned. I cannot
put my mind at ease.
I have trouble
sleeping.
My daughter Virginia
reads the prescribed
penitential psalms.
What is Christianity
that it can be threatened?
The sun is turning on its axis.
I am old.
I am going blind.

IV. Caroline Herschel (1750-1848)

for Carol

William is away, and I am minding
the heavens. I have discovered
eight new comets and three nebulae
never before seen by man,
and I am preparing an Index to
Flamsteed's observations, together with
a catalogue of 560 stars omitted from
the British Catalogue, plus a list of errata
in that publication. William says

I have a way with numbers, so I handle
all the necessary reductions and
calculations. I also plan
every night's observation
schedule, for he says my intuition
helps me turn the telescope to discover
star cluster after star cluster.

I have helped him polish the mirrors
and lenses of our new telescope. It is
the largest in existence. Can you imagine
the thrill of turning it to some new
corner of the heavens to see
something never before seen
from earth? I actually like

that he is busy with the Royal Society
and his club, for when I finish my other work
I can spend all night sweeping
the heavens.

Sometimes when I am alone
in the dark, and the universe reveals
yet another secret, I say the names
of my long lost sisters, forgotten
in the books that record
our science—

> Aglaonice of Thessaly,
> Hyptia,
> Hildegard,
> Catherina Hevelius,
> Maria Agnesi

—as if the stars themselves could

remember. Did you know that Hildegard
proposed a heliocentric universe
300 years before Copernicus? that she
wrote of universal gravitation 500 years
before Newton? But who would listen
to her? She was just a nun, a woman.
What is our age, if that age was dark?

As for my name, it will also be
forgotten, but I am not accused
of being a sorceress, like Aglaonice,
and the Christians do not threaten to
drag me to church, to murder me, like they did
Hyptia of Alexandria, the eloquent, young
woman who devised the instruments
used to accurately measure the position
and motion of
heavenly bodies.

However long we live, life is short, so I
work. And however important man becomes,
he is nothing compared to the stars.
There are secrets, dear sister, and it is
for us to reveal them. Your name, like mine,
is a song. Write soon,
Caroline

V. Albert Einstein (1879-1955)

Yes, I have written
the President. I have told him
that if there is a nuclear war, the Fourth
World War will be fought
with sticks and stones.

Words do hurt me,
and there is no change in my heart
condition, but I am trying to complete
my unified field
theory. I cannot believe
that God is playing roulette
with the world. The mystery
must be locked up
in the elemental infrastructures.

Forgive me for using
scrap paper. The other day
when my wife and I
were being shown
the huge reflecting telescope
at Mount Wilson observatory,
she asked why the instruments
were so large.

On being told
that they were trying to discern
the shape and makeup of the whole
universe, she said:
My husband does that
on the back
of an old
envelope.

RICHARD FEYNMAN

1918-1988

QED: The Strange Theory of Light and Matter

1985

A native of New York City, Richard Feynman was educated at the Massachusetts Institute of Technology and Princeton University. While studying for his doctorate in physics at Princeton, he worked on the Manhattan Project both in Princeton and in Los Alamos where he helped to develop a theory of how to separate Uranium 235 from Uranium 238. After teaching five years at Cornell University, he became a professor at the California Institute of Technology, which became his lifelong academic home. Recognized as one of the most accomplished physicists of the twentieth century, Feynman was known as a charismatic lecturer for his ability to teach the most complex principles to his students.

For his research into quantum electrodynamics and particle physics, he was awarded the Nobel Prize in physics in 1965. He introduced diagrams, called Feynman diagrams, "graphic analogues of the mathematical expressions needed to describe the behavior of systems of interacting particles."[1]

---

[1] http://www-history.mcs.st-andrews.ac.uk/history/Biographies/ Feynman.html

A member of many distinguished societies in the United States and Great Britain, he also won many awards for his research. He was a member of the Rogers Commission, which studied the *Challenger* disaster. Annoyed at the vagueness of answers to the panel's questions, he tossed an O-ring into a class of ice water to show that even at the temperature of ice water, much less at the colder temperatures of outer space, the O-ring would not expand, thus demonstrating the role of the rings in the explosion of the space shuttle. His thoughts and concerns about his work with the Commission are chronicled in *What Do You Care What Other People Think?: Further Adventures of a Curious Character.*

Feynman's conversational style of teaching is illustrated by his 1985 book, *Surely You're Joking, Mr. Feynman,* which became a best seller. His book, *QED: The Strange Theory of Light and Matter,* was also extremely popular. It comprises four lectures on quantum electrodynamics (QED). In the following excerpt, Feynman explains the parameters of QED, or the interactions between electrons and light, in a surprisingly accessible way.

Sources:

"Richard Phillips Feynman." 2003. Retrieved February 2, 2007, from http://www.nobel-winners.com/Physics/richard_phillips_feynman.html.

"Richard Phillips Feynman." 2002. School of Mathematics and Statistics, University of St. Andrews, Scotland. Retrieved February 2, 2007, from http://www-history.mcs.st-andrews.ac.uk/history/Biographies/Feynman.html.

Selection From:

Feynman, Richard P. 1985. *QED: The Strange Theory of Light and Matter.* Princeton: Princeton University Press. 3-35.

Alix Mautner was very curious about physics and often asked me to explain things to her. I would do all right, just as I do with a group of students at Caltech that come to me for an hour on Thursdays, but eventually I'd fail at what is to me the most interesting part: We would always get hung up on the crazy ideas of quantum mechanics. I told her I couldn't explain these ideas in an hour or an evening—it would take a long time—but I promised her that someday I'd prepare a set of lectures on the subject.

I prepared some lectures, and I went to New Zealand to try them out—because New Zealand is far enough away that if they weren't successful, it would be all right! Well, the people in New Zealand thought they were okay, so I guess they're okay—at least for New Zealand! So here are the lectures I really prepared for Alix, but unfortunately I can't tell them to her directly, now.

What I'd like to talk about is a part of physics that is *known*, rather than a part that is unknown. People are always asking for the latest developments in the unification of this theory with that theory, and they don't give us a chance to tell them anything about one of the theories that we know pretty well. They always want to know things that we don't know. So, rather than confound you with a lot of half-cooked, partially analyzed theories, I would like to tell you about a subject that has been very thoroughly analyzed. I love this area of physics and I think it's wonderful: it is called quantum electrodynamics, or QED for short.

My main purpose in these lectures is to describe as accurately as I can the strange theory of light and matter—or more specifically, the interaction of light and electrons. It's going to take a long time to explain all the things I want to. However, there are four lectures, so I'm going to take my time, and we will get everything all right.

Physics has a history of synthesizing many phenomena into a few theories. For instance, in the early days there were phenomena of motion and phenomena of heat; there were phenomena of sound, of light, and of gravity. But it was soon discovered, after Sir Isaac Newton explained the laws of motion, that some of these apparently different things were aspects of the same thing. For example, the phenomena of sound could be completely understood as the motion of atoms in the air.

So sound was no longer considered something in addition to motion. It was also discovered that heat phenomena are easily understandable from the laws of motion. In this way, great globs of physics theory were synthesized into a simplified theory. The theory of gravitation, on the other hand, was not understandable from the laws of motion, and even today it stands isolated from the other theories. Gravitation is, so far, not understandable in terms of other phenomena.

After the synthesis of the phenomena of motion, sound, and heat, there was the discovery of a number of phenomena that we call electrical and magnetic. In 1873 these phenomena were synthesized with the phenomena of light and optics into a single theory by James Clerk Maxwell, who proposed that light is an electromagnetic wave. So at that stage, there were the laws of motion, the laws of electricity and magnetism, and the laws of gravity.

Around 1900 a theory was developed to explain what matter was. It was called the electron theory of matter, and it said that there were little charged particles inside of atoms. This theory evolved gradually to include a heavy nucleus with electrons going around it.

Attempts to understand the motion of the electrons going around the nucleus by using mechanical laws—analogous to the way Newton used the laws of motion to figure out how the earth went around the sun—were a real failure: all kinds of predictions came out wrong. (Incidentally, the theory of relativity, which you all understand to be a great revolution in physics, was also developed at about that time. But compared to this discovery that Newton's laws of motion were quite wrong in atoms, the theory of relativity was only a minor modification.) Working out another system to replace Newton's laws took a long time because phenomena at the atomic level were quite strange. One had to lose one's common sense in order to perceive what was happening at the atomic level. Finally, in 1926, an "uncommon-sensy" theory was developed to explain the "new type of behavior" of electrons in matter. It looked cockeyed, but in reality it was not: it was called the theory of quantum mechanics. The word "quantum" refers to this peculiar aspect of nature that goes against common sense. It is this aspect that I am going to tell you about.

The theory of quantum mechanics also explained all kinds of details, such as why an oxygen atom combines with two hydrogen atoms to make water, and so on. Quantum mechanics thus supplied the theory behind chemistry. So, fundamental theoretical chemistry is really physics.

Because the theory of quantum mechanics could explain all of chemistry and the various properties of substances, it was a tremendous success. But still there was the problem of the interaction of light and matter. That is, Maxwell's theory of electricity and magnetism had to be changed to be in accord with the new principles of quantum mechanics that had been developed. So a new theory, the quantum theory of the interaction of light and matter, which is called by the horrible name "quantum electrodynamics," was finally developed by a number of physicists in 1929.

But the theory was troubled. If you calculated something roughly, it would give a reasonable answer. But if you tried to compute it more accurately, you would find that the correction you thought was going to be small (the next term in a series for example) was in fact very large—in fact it was *infinity*! So it turned out you couldn't really compute *anything* beyond a certain accuracy.

By the way, what I have just outlined is what I call a "physicist's history of physics," which is never correct. What I am telling you is a sort of conventionalized myth-story that the physicists tell to their students, and those students tell to their students, and is not necessarily related to the actual historical development, which I do not really know!

At any rate, to continue with this "history," Paul Dirac using the theory of relativity, made a relativistic theory of the electron that did not take into account the effects of the electron's interaction with light. Dirac's theory said that an electron had a magnetic moment—something like the force of a little magnet—that had the strength of exactly 1 in certain units. Then in about 1948 it was discovered that in experiments that the actual number was closer to 1.00118 (with an uncertainty of about 3 on the last digit). It was known, of course, that electrons interact with light, so some small correction was expected. It was also expected that this correction would be understandable from the new theory of quantum electrodynamics. But when it was calculated, instead of 1.00118 the result was infinity—which is wrong, experimentally!

Well, this problem of how to calculate things in quantum electrodynamics was straightened out by Julian Schwinger, Sin-Itiro Tomonga, and myself in about 1948. Schwinger was the first to calculate this correction using a new "shell game"; his theoretical value was around 1.00116, which was close enough to the experimental number to show that we were on the right track. At last, we had a quantum theory of electricity and magnetism with which we could calculate! This is the theory that I am going to describe to you.

The theory of quantum electrodynamics has now lasted for more than fifty years, and has been tested more and more accurately over a wider and wider range of conditions. At the present time I can proudly say that there is *no significant difference* between experiment and theory!

Just to give you an idea of how the theory has been put through the wringer, I'll give you some recent experiments have Dirac's number at 1.00115965221 (with an uncertainty of about 4 in the last digit); the theory puts it at 1.0011595246 (with an uncertainty of about five times as much). To give you a feeling for the accuracy of these numbers, it comes out something like this: If you were to measure the distance from Los Angeles to New York to this accuracy, it would be exact to the thickness of a human hair. That's how delicately quantum electrodynamics has, in the past fifty years, been checked—both theoretically and experimentally. By the way, I have chosen only one number to show you. There are other things in quantum electrodynamics that have been measured with comparable accuracy, which also agree very well. Things have been checked at distance scales that range from one hundred times the size of the earth down to one-hundredth the size of an atomic nucleus. These numbers are meant to intimidate you into believing that the theory is probably not too far off! Before we're through, I'll describe how these calculations are made.

I would like to again impress you with the vast range of phenomena that the theory of quantum electrodynamics describes: It's easier to say it backwards: the theory describes *all* the phenomena of the physical world except the gravitational effect, the thing that holds you in your seats (actually, that's a combination of gravity and politeness, I think) and radioactive phenomena, which involve nuclei shifting in their energy levels. So if we leave out gravity and radioactivity (more properly, nuclear physics), what have we got left? Gasoline burning in automobiles, foam and bubbles, the hardness of salt or copper, the stiffness of steel. In fact, biologists are trying to interpret as much as they can about life in terms of chemistry, and as I already explained, the theory behind chemistry is quantum electrodynamics.

I must clarify something: When I say that all the phenomena of the physical world can be explained by this theory, we don't really know that. Most phenomena we are familiar with involve such *tremendous* numbers of electrons that it's hard for our poor minds to follow that complexity. In such situations, we can use the theory to figure roughly what ought to happen and that *is* what happens, roughly, in those circumstances. But if we arrange in the laboratory an experiment involving just a *few* electrons in *simple* circumstances, then we can calculate what might happen very accurately, and we can measure it very accurately, too. Whenever we do such experiments, the theory of quantum electrodynamics works very well.

We physicists are always checking to see if there is something the matter with the theory. That's the game, because if there *is* something the matter, it's interesting! But so far, we have found nothing wrong with the theory of quantum electrodynamics. It is, therefore, I would say, the jewel of physics—our proudest possession.

The theory of quantum electrodynamics is also the prototype for new theories that attempt to explain nuclear phenomena, the things that go on inside the nuclei of atoms. If one were to think of the physical world as a stage, then the actors would be not only electrons, which are outside the nucleus in atoms, but also quarks and gluons and so forth—dozens of kinds of particles—inside the nucleus. And though these "actors" appear quite different from one another, they all act in a certain style—a strange and peculiar style—the "quantum" style. At the end, I'll tell you a little bit about the nuclear particles. In the meantime, I'm only going to tell you about photons—particles of light—and electrons, to keep it simple. Because it's the way they act that is important, and the way they act is very interesting.

So now you know what I'm going to talk about. The next question is, will you *understand* what I'm going to tell you? Everybody who comes to a scientific lecture knows they are not going to understand it, but maybe the lecturer has a nice, colored tie to look at. Not in this case! (Feynman is not wearing a tie.)

What I am going to tell you about is what we teach our physics students in the third or fourth year of graduate school—and you think I'm going to explain it to you so you can understand it? No, you're not going to be able to understand it. Why, then, am I going to bother you with all this? Why are you going to sit here all this time, when you won't be able to understand what I am going to say? It is my task to convince you *not* to turn away because you don't understand it. You see, my physics students don't understand it either. That is because *I* don't understand it. Nobody does.

I'd like to talk a little bit about understanding. When we have a lecture, there are many reasons why you might not understand the speaker. One is, his language is bad—he doesn't say what he means to say, or he says it upside down—and it's hard to understand. That's a rather trivial matter, and I'll try my best to avoid too much of my New York accent.

Another possibility, especially if the lecturer is a physicist, is that he uses ordinary words in a funny way. Physicists often use ordinary words such as "work" or "action" or "energy" or even, as you shall see, "light" for some technical purpose. Thus, when I talk about "work" in physics, I don't mean the same thing as when I talk about "work" on the street. During this lecture I might use one of those words without noticing that it is being used in this unusual way. I'll try my best to catch myself—that's my job—but it is an error that is easy to make.

The next reason that you might think you do not understand what I am telling you is, while I am describing to you *how* Nature works, you won't understand *why* Nature works that way. But you see, nobody understands that. I can't explain why Nature behaves in this peculiar way.

Finally, there is this possibility: after I tell you something, you just can't believe it. You can't accept it. You don't like it. A little screen comes down and you don't listen anymore. I'm going to describe to you how Nature is—and if you don't like it, that's going to get in the way of your understanding it. It's a problem that physicists have learned to deal with: They've learned to realize that whether they like a theory or they don't like a theory is not the essential question. Rather, it is whether or not the theory gives predictions that agree with experiment. It is not a question of whether a theory is philosophically delightful, or easy to understand, or perfectly reasonable from the point of view of common sense. The theory of quantum electrodynamics describes Nature as absurd from the point of view of common sense. And it agrees fully with experiment. So I hope you can accept Nature as She is—absurd.

I'm going to have fun telling you about this absurdity, because I find it delightful. Please don't turn yourself off because you can't believe Nature is so strange. Just hear me all out, and I hope you'll be as delighted as I am when we're through.

How am I going to explain to you the things I don't explain to my students until they are third-year graduate students? Let me explain it by analogy. The Maya Indians were interested in the rising and setting of Venus as a morning "star" and as an evening "star"—they were very interested in when it would appear. After some years of observation, they noted that five cycles of Venus were very nearly equal to eight of their "nominal years" of 365 days (they were aware that the true year of seasons was different and they made calculations of that also). To make calculations, the Maya had invented a system of bars and dots to represent numbers (including zero), and had rules by which to calculate and predict not only the risings and settings of Venus, but other celestial phenomena, such as lunar eclipses.

In those days, only a few Maya priests could do such elaborate calculations. Now, suppose we were to ask one of them how to do just one step in the process of predicting when Venus will next rise as a morning star—subtracting two numbers. And let's assume that, unlike today, we had not gone to school and did not know how to subtract. How would the priest explain to us what subtraction is?

He could either teach us the numbers represented by the bars and dots and the rules for "subtracting" them, or he could tell us what he was really doing: "Suppose we want to subtract 236 from 584. First, count out 584 beans and put them in a pot. Then take out 236 beans and put them to one side. Finally, count the beans left in the pot. That number is the result of subtracting 236 from 584."

You might say, "My Quetzalcoatl! What *tedium*—counting beans, putting them in, taking them out—what a job!" To which the priest would reply, "That's why we have the rules for the bars and dots. The rules are tricky, but they are a much more efficient way of getting the answer than by counting beans. The important thing is, it makes no difference as far as the *answer* is concerned: we can predict the appearance of Venus by counting beans (which is slow, but easy to understand) or by using the tricky rules (which is much faster, but you must spend years in school to learn them)."

To understand *how* subtraction works—as long as you don't have to actually carry it out—is really not so difficult. That's my position: I'm going to explain to you what the physicists are *doing* when they are predicting how Nature will behave, but I'm not going to teach you any tricks so you can do it *efficiently*. You will discover that in order to make any reasonable predictions with this new scheme of quantum electrodynamics, you would have to make an awful lot of little arrows on a piece of paper. It takes seven years—four undergraduate and three graduate—to train our physics students to do that in a tricky, efficient way. That's where we are going to skip seven years of education in physics: By explaining quantum electrodynamics to you in terms of what we are *really doing*, I hope you will be able to understand it better than do some of the students!

Taking the example of the Maya one step further, we, could ask the priest *why* five cycles of Venus nearly equal 2,920 days, or eight years. There would be all kinds of theories about *why*, such as, "20 is an important number in our counting system, and if you divide 2,920 by 20, you get 146, which is one more than a number that can be represented by the sum of two squares in two different ways," and so forth. But that theory would have nothing to do with Venus, really. In modern times, we have found that theories of this kind are not useful. So again, we are not going to deal with *why* Nature behaves in the peculiar way that She does; there are no good theories to explain that.

What I have done so far is to get you into the right mood to listen to me. Otherwise, we have no chance. So now we're off, ready to go!

We begin with light. When Newton started looking at light, the first thing he found was that white light is a mixture of colors. He separated white light with a prism into various colors, but when he put light of one color—red, for instance—through another prism, he found it could not be separated further. So Newton found that white light is a mixture of different colors, each of which is pure in the sense that it can't be separated further.

(In fact, a particular color of light can be split one more time in a different way, according to its so-called "polarization." This aspect of light is not vital to understanding the character of quantum electrodynamics, so for the sake of simplicity I will leave it out—at the expense of not giving you an absolutely complete description of the theory. This slight simplification will not remove, in any way, any real understanding of what I will be talking about. Still, I must be careful to mention all of the things I leave out.)

When I say "light" in these lectures, I don't mean simply the light we can see, from red to blue. It turns out that visible light is just a part of a long scale that's analogous to a musical scale in which there are notes higher than you can hear and other notes lower than you can hear. The scale of light can be described by numbers—called the frequency—and as the numbers get higher, the light goes from red to blue to violet to ultraviolet. We can't see ultraviolet light, but it can affect photographic plates. It's still light—only the number is different. (We shouldn't be so provincial: what we can detect directly with our own instrument, the eye, isn't the only thing in the world!) If we continue simply to change the number, we go out into X-rays, gamma rays, and so on. If we change the number in the other direction, we go from blue to red to infrared (heat) waves, then television waves, and radio waves. For me, all of that is "light." I'm going to use just red light for most of my examples, but the theory of quantum electrodynamics extends over the entire range I have described, and is the theory behind all these various phenomena.

Newton thought that light was made up of particles—he called them "corpuscles"—and he was right (but the reasoning that he used to come to that decision was erroneous). We know that light is made of particles because we can take a very sensitive instrument that makes clicks when light shines on it, and if the light gets dimmer, the clicks remain just as loud—there are just fewer of them. Thus light is something like raindrops—each little lump of light is called a photon—and if the light is all one color, all the "raindrops" are the same size.

The human eye is a very good instrument: it takes only about five or six photons to activate a nerve cell and send a message to the brain. If we were evolved a little further so we could see ten times more sensitively, we wouldn't have to have this discussion—we would all have seen very dim light of one color as a series of intermittent little flashes of equal intensity.

You might wonder how it is possible to detect a single photon. One instrument that can do this is called a photomultiplier, and I'll describe briefly how it works: When a photon hits the metal plate A at the bottom (see Figure 1), it causes an electron to break loose from one of the atoms in the plate. The free electron is strongly attracted to plate B (which has a positive charge on it) and hits it with enough force to break loose three or four electrons. Each of the electrons knocked out of plate B is attracted to plate C (which is also charged), and their collision with plate C knocks loose even more electrons. This process is repeated ten or twelve times, until billions of electrons, enough to make a sizable electric current, hit the last plate, L. This current can be amplified by a regular amplifier and sent through a speaker to make audible clicks. Each time a photon of a given color hits the photomultiplier, a click of uniform loudness is heard.

If you put a whole lot of photomultipliers around and let some very dim light shine in various directions, the light goes into one multiplier or another and makes a click of full intensity. It is all or nothing: if one photomultiplier goes off at a given moment, none of the others goes off at the same moment (except in the rare instance that two photons happened to leave the light source at the same time). There is no splitting of light into "half particles" that go different places.

FIGURE 1. *A photomultiplier can detect a single photon. When a photon strikes plate A, an electron is knocked loose and attracted to positively charged plate B, knocking more electrons loose. This process continues until billions of electrons strike the last plate, L, and produce an electric current, which is amplified by a regular amplifier. If a speaker is connected to the amplifier, clicks of uniform loudness are heard each time a photon of a given color hits plate A.*

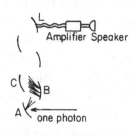

I want to emphasize that light comes in this form—particles. It is very important to know that light behaves like particles, especially for those of you who have gone to school, where you were probably told something about light behaving like waves. I'm telling you the way it *does* behave—like particles.

You might say that it's just the photomultiplier that detects light as particles, but no, every instrument that has been designed to be sensitive enough to detect weak light has always ended up discovering the same thing: light is made of particles.

I am going to assume that you are familiar with the properties of light in everyday circumstances—things like, light goes in straight lines; it bends when it goes into water; when it is reflected from a surface like a mirror, the angle at which the light hits the surface is equal to the angle at which it leaves the surface; light can be separated into colors; you can see beautiful colors on a mud puddle when there is a little bit of oil on it; a lens focuses light, and so on. I am going to use these phenomena that you are familiar with in order to illustrate the truly strange behavior of light; I am going to explain these familiar phenomena in terms of the theory of quantum electrodynamics. I told you about the photomultiplier in order to illustrate an essential phenomenon that you may not have been familiar with—that light is made of particles—but by now, I hope you are familiar with that, too!

Now, I think you are all familiar with the phenomenon that light is partly reflected from some surfaces, such as water. Many are the romantic paintings of moonlight reflecting from a lake (and many are the times you got yourself in trouble *because* of moonlight reflecting from a lake!). When you look down into water you can see what's below the surface (especially in the daytime), but you can also see a reflection from the surface. Glass is another example: if you have a lamp on in the room and you're looking out through a window during the daytime, you can see things outside through the glass as well as a dim reflection of the lamp in the room. So light is partially reflected from the surface of glass.

Before I go on, I want you to be aware of a simplification I am going to make that I will correct later on: When I talk about the partial reflection of light by glass, I am going to pretend that the light is reflected by only the surface of the glass. In reality, a piece of glass is a terrible monster of complexity—huge numbers of electrons are jiggling about. When a photon comes down, it interacts with electrons throughout the glass, not just on the surface. The photon and electrons do some kind of dance, the net result of which is the same as if the photon hit only the surface. So let me make that simplification for a while. Later on, I'll show you what actually happens inside the glass so you can understand why the result is the same.

Now I'd like to describe an experiment, and tell you its surprising results. In this experiment some photons of the same color—let's say, red light—are emitted from a light source (see Fig. 2) down toward a block of glass. A photomultiplier is placed at A, above the glass, to catch any photons that are reflected by the front surface. To measure how many photons get past the front surface, another photomultiplier is placed at B, inside the glass. Never mind the obvious difficulties of putting a photomultiplier inside a block of glass; what are the results of this experiment?

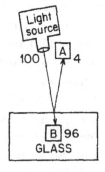

FIGURE 2. *An experiment to measure the partial reflection of light by a single surface of glass. For every 100 photons that leave the light source, 4 are reflected by the front surface and end up in the photomultiplier at A, while the other 96 are transmitted by the front surface and end up in the photomultiplier at B.*

For every 100 photons that go down toward the glass, an average of 4 arrive at A and 96 arrive at B. So "partial reflection" in this case means that 4% of the photons are reflected by the front surface of the glass, while the other 96% are transmitted. *Already* we are in great difficulty: how can light be *partly* reflected? Each photon ends up at A or B—how does the photon "make up its mind" whether it should go to A or B?

(Audience laughs.) That may sound like a joke, but we can't just laugh; we're going to have to explain that in terms of a theory! Partial reflection is already a deep mystery, and it was a very difficult problem for Newton.

There are several possible theories that you could make up to account for the partial reflection of light by glass. One of them is that 96% of the surface of the glass is "holes" that let the light through, while the other 4% of the surface is covered by small "spots" of reflective material (see Fig. 3). Newton realized that this is not a possible explanation.[2] In just a moment we will encounter a strange feature of partial reflection that will drive you crazy if you try to stick to a theory of "holes and spots"—or to any other reasonable theory!

Another possible theory is that the photons have some kind of internal mechanism—"wheels" and "gears" inside that are turning in some way—so that when a photon is "aimed" just right, it goes through the glass, and when it's not aimed right, it reflects. We can check this theory by trying to filter out the photons that are not aimed right by putting a few extra layers of glass between the source and the first layer of glass. After going through the filters, the photons reaching

---

[2] How did he know? Newton was a very great man: he wrote, "Because I can polish glass." You might wonder, how the heck could he tell that because you can polish glass, it can't be holes and spots? Newton polished his own lenses and mirrors, and he knew what he was doing with polishing: he was making scratches on the surface of a piece of glass with powders of increasing fineness. As the scratches become finer and finer, the surface of the glass changes its appearance from a dull grey (because the light is scattered by the large scratches), to a transparent clarity (because the extremely fine scratches let the light through). Thus he saw that it is impossible to accept the proposition that light can be affected by very small irregularities such as scratches or holes and spots; in fact, he found the contrary to be true. The finest scratches and therefore equally small spots do not affect the light. So the holes and spots theory is no good.

the glass should *all* be aimed right, and none of them should reflect. The trouble with that theory is, it doesn't agree with experiment: even after going through many layers of glass, 4% of the photons reaching a given surface reflect off it.

Try as we might to invent a reasonable theory that can explain how a photon "makes up its mind" whether to go through glass or bounce back, it is impossible to predict which way a given photon will go. Philosophers have said that if the same circumstances don't always produce the same results, predictions are impossible and science will collapse. Here is a circumstance—identical photons are always coming down in the same direction to the same piece of glass—that produces different results. We cannot predict whether a given photon will arrive at A or B. All we can predict is that out of 100 photons that come down, an average of 4 will be reflected by the front surface. Does this mean that physics, a science of great exactitude, has been reduced to calculating only the *probability* of an event, and not predicting exactly what will happen? Yes. That's a retreat, but that's the way it is: Nature permits us to calculate only probabilities. Yet science has not collapsed.

While partial reflection by a single surface is a deep mystery and a difficult problem, partial reflection by two or more surfaces is absolutely mind-boggling. Let me show you why. We'll do a second experiment, in which we will measure the partial reflection of light by two surfaces. We replace the block of glass with a thin sheet of glass—its two surfaces are exactly parallel to each other—and we place the photomultiplier below the sheet of glass, in line with the light source. This time, photons can reflect from either the front surface or the back surface to end up at A; all the others will end up at B (see Fig. 4). We might expect the front surface to reflect 4% of the light and the back surface to reflect 4% of the remaining 96%, making a total of about 8%. So we should find that out of every 100 photons that leave the light source, about 8 arrive at A.

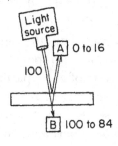

FIGURE 4. *An experiment to measure the partial reflection of light by two surfaces of glass. Photons can get to the photomultiplier at A by reflecting off either the front surface or the back surface of the sheet of glass; alternatively, they could go through both surfaces and end up hitting the photomultiplier at B. Depending on the thickness of the glass, 0 to 16 photons out of every 100 get to the photomultiplier at A. These results pose difficulties for any reasonable theory, including the one in Figure 3. It appears that partial reflection can be "turned off" or "amplified" by the presence of an additional surface.*

What actually happens under these carefully controlled experimental conditions is, the number of photons arriving at A is rarely 8 out of 100. With some sheets of glass, we consistently get a reading of 15 or 16 photons—twice our expected result! With other sheets of glass, we consistently get only 1 or 2 photons. Other sheets of glass give a partial reflection of 10%; some eliminate partial reflection altogether! What can account for these crazy results? After checking the various sheets of glass for quality and uniformity, we discover that they differ only in their thickness.

To test the idea that the amount of light reflected by two surfaces depends on the thickness of the glass, let's do a series of experiments: Starting out with the thinnest possible layer of glass, we'll count how many photons hit the photomultiplier at A each time 100 photons leave the light source. Then we'll replace the layer of glass with a slightly thicker one and make new counts. After repeating this process a few dozen times, what are the results?

With the thinnest possible layer of glass, we find that the number of photons arriving at A is nearly always zero—sometimes it's 1. When we replace this thinnest layer with a slightly thicker one, we find that the amount of light reflected is higher—closer to the expected 8%. After a few more replacements the count of photons arriving at A increases past the 8% mark. As we continue to substitute still thicker and thicker layers of glass, the amount of light reflected by the two surfaces reaches a maximum of 16%, and then goes down, through 8%, back to zero—if the layer of glass is just the right thickness, there is no reflection at all. (Do *that* with spots!)

With gradually thicker and thicker layers of glass, partial reflection again increases to 16% and returns to zero—a cycle that repeats itself again and again (see Fig. 5). Newton discovered these oscillations and did one experiment that could be correctly interpreted only if the oscillations continued for at least 34,000 cycles! Today, with lasers (which produce a very pure, monochromatic light), we can see this cycle still going strong after more than 100,000,000 repetitions—which corresponds to glass that is more than 50 meters thick.

So it turns out that our prediction of 8% is right as an overall average (since the actual amount varies in a regular pattern from zero to 16%), but it's exactly right only twice each cycle—like a stopped clock (which is right twice a day). How can we explain this strange feature of partial reflection that depends on the thickness of the glass? How can the front surface reflect 4% of the light (as confirmed in our first experiment) when, by putting a second surface at just the right distance below, we can somehow "turn off" the reflection? And by placing that second surface at a slightly different depth, we can "amplify" the reflection up to 16%! Can it be that the back surface exerts some kind of influence or effect on the ability of the front surface to reflect light? What if we put in a *third* surface?

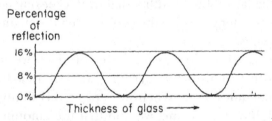

FIGURE 5. *The results of an experiment carefully measuring the relationship between the thickness of a sheet of glass and partial reflection demonstrate a phenomenon called "interference." As the thickness of the glass increases, partial reflection goes through a repeating cycle of zero to 16%, with no signs of dying out.*

With a third surface, or any number of subsequent surfaces, the amount of partial reflection is again changed. We find ourselves chasing down through surface after surface with this theory, wondering if we have finally reached the last surface. Does a photon have to do that in order to "decide" whether to reflect off the front surface?

Newton made some ingenious arguments concerning this problem,[3] but he realized, in the end, that he had not yet developed a satisfactory theory.

---

[3] It is very fortunate for us that Newton convinced himself that light is "corpuscles," because we can see what a fresh and intelligent mind looking at this phenomenon of partial reflection by two or more surfaces has to go through to try to explain it. (Those who believed that light was waves never had to wrestle with it.) Newton argued as follows: Although light appears to be reflected from the first surface, it cannot be reflected from that surface. If it were, then how could light reflected from the first surface be captured again when the thickness is such that there was supposed to be no reflection at all? Then light must be reflected from the second surface. But to account for the fact that the thickness of the glass determines the amount of partial reflection, Newton proposed this idea: Light striking the first surface sets off a kind of wave or field that travels along with the light and predisposes it to reflect or not reflect off the second surface. He called this process "fits of easy reflection or easy transmission" that occur in cycles, depending on the thickness of the glass.

There are two difficulties with this idea: the first is the effect of additional surfaces—each new surface affects the reflection—which I described in the text. The other problem is that light certainly reflects off a lake, which doesn't have a second surface, so light *must* be reflecting off the front surface. In the case of single surfaces, Newton said that light had a predisposition to reflect. Can we have a theory in which the light knows what kind of surface it is hitting, and whether it is the only surface?

For many years after Newton, partial reflection by two surfaces was happily explained by a theory of waves,[4] but when experiments were made with very weak light hitting photomultipliers, the wave theory collapsed: as the light got dimmer and dimmer, the photomultipliers kept making full-sized clicks—there were just fewer of them. Light behaved as particles.

The situation today is, we haven't got a good model to explain partial reflection by two surfaces; we just calculate

---

Newton didn't emphasize these difficulties with his theory of "fits of reflection and transmission," even though it is clear that he knew his theory was not satisfactory. In Newton's time, difficulties with a theory were dealt with briefly and glossed over—a different style from what we are used to in science today, where we point out the places where our own theory doesn't fit the observations of experiment. I'm not trying to say anything against Newton; I just want to say something in favor of how we communicate with each other in science today.

[4] This idea made use of the fact that waves can combine or cancel out, and the calculations based on this model matched the results of Newton's experiments, as well as those done for hundreds of years afterwards. But when instruments were developed that were sensitive enough to detect a single photon, the wave theory predicted that the "clicks" of the photomultiplier would get softer and softer, whereas they stayed at full strength—they just occurred less and less often. No reasonable model could explain this fact, so there was a period for a while in which you had to be clever: You had to know which experiment you were analyzing in order to tell if light was waves or particles. This state of confusion was called the "wave-particle duality" of light, and it was jokingly said by someone that light was waves on Mondays, Wednesdays, and Fridays; it was particles on Tuesdays, Thursdays, and Saturdays, and on Sundays, we think about it! It is the purpose of these lectures to tell you how this puzzle was finally "resolved."

the probability that a particular photomultiplier will be hit by a photon reflected from a sheet of glass. I have chosen this calculation as our first example of the method provided by the theory of quantum electrodynamics. I am going to show you "how we count the beans"—what the physicists do to get the right answer. I am not going to explain how the photons actually "decide" whether to bounce back or go through; that is not known. (Probably the question has no meaning.) I will only show you how to calculate the correct *probability* that light will be reflected from glass of a given thickness, because that's the only thing physicists know how to do! What we do to get the answer to *this* problem is analogous to the things we have to do to get the answer to *every other* problem explained by quantum electrodynamics.

You will have to brace yourselves for this—not because it is difficult to understand, but because it is absolutely ridiculous: All we do is draw little arrows on a piece of paper—that's all!

Now, what does an arrow have to do with the chance that a particular event will happen? According to the rules of "how we count the beans," the probability of an event is equal to the square of the length of the arrow. For example, in our first experiment (when we were measuring partial reflection by the front surface only), the probability that a photon would arrive at the photomultiplier at A was 4%. That corresponds to an arrow whose length is 0.2, because 0.2 squared is 0.04 (see Fig. 6).

In our second experiment (when we were replacing thin sheets of glass with slightly thicker ones), photons bouncing off either the front surface or the back surface arrived at A. How do we draw an arrow to represent this situation? The length of the arrow must range from zero to 0.4 to represent probabilities of zero to 16%, depending on the thickness of the glass (see Fig. 7).

FIGURE 6. *The strange feature of partial reflection by two surfaces has forced physicists away from making absolute predictions to merely calculating the probability of an event. Quantum electrodynamics provides a method for doing this—drawing little arrows on a piece of paper. The probability of an event is represented by the area of the square on an arrow. For example, an arrow representing a probability of 0.04 (4%) has a length of 0.2.*

We start by considering the various ways that a photon could get from the source to the photomultiplier at A. Since I am making this simplification that the light bounces off either the front surface or the back surface, there are two possible ways a photon could get to A. What we do in this case is to draw *two* arrows—one for each way the event can happen—and then combine them into a "final arrow" whose square represents the probability of the event. If there had been three different ways the event could have happened, we would have drawn three separate arrows before combining them.

FIGURE 6. *The strange feature of partial reflection by two surfaces has forced physicists away from making absolute predictions to merely calculating the probability of an event. Quantum electrodynamics provides a method for doing this—drawing little arrows on a piece of paper. The probability of an event is represented by the area of the square on an arrow. For example, an arrow representing a probability of 0.04 (4%) has a length of 0.2.*

Now, let me show you how we combine arrows. Let's say we want to combine arrow x with arrow y (see Fig. 8). All we have to do is put the head of x against the tail of y (without changing the direction of either one), and draw the final arrow from the tail of x to the head of y. That's all there is to it. We can combine any number of arrows in this manner (technically, it's called "adding arrows"). Each arrow tells you how far, and in what direction, to move in a dance. The final arrow tells you what *single* move to make to end up in the same place (see Fig. 9).

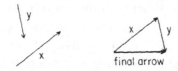

FIGURE 8. *Arrows that represent each possible way an event could happen are drawn and then combined ("added") in the following manner: Attach the head of one arrow to the tail of another—without changing the direction of either one—and draw a "final arrow" from the tail of the first arrow to the head of the last one.*

Now, what are the specific rules that determine the length and direction of each arrow that we combine in order to make the final arrow? In this particular case, we will be combining two arrows—one representing the reflection from the *front* surface of the glass, and the other representing the reflection from the *back* surface.

Let's take the length first. As we saw in the first experiment (where we can put the photomultiplier inside the glass), the front surface reflects about 4% of the photons that come down. That means the "front reflection" arrow has a length of 0.2. The back surface of the glass also reflects 4%, so the "back reflection" arrow's length is also 0.2.

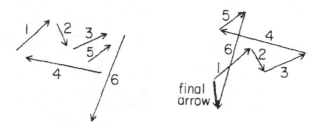

FIGURE 9. *Any number of arrows can be added in the manner described in Figure 8.*

To determine the direction of each arrow, let's imagine that we have a stopwatch that can time a photon as it moves. This imaginary stopwatch has a single hand that turns around very, very rapidly. When a photon leaves the source, we start the stopwatch. As long as the photon moves, the stopwatch hand turns; when the photon ends up at the photomultiplier, we stop the watch. The hand ends up pointing in a certain direction. That is the direction we will draw the arrow.

We need one more rule in order to compute the answer correctly: When we are considering the path of a photon bouncing off the *front* surface of the glass, we reverse the direction of the arrow. In other words, whereas we draw the *back* reflection arrow pointing in the same direction as the stopwatch hand, we draw the *front* reflection arrow in the *opposite* direction.

Now, let's draw the arrows for the case of light reflecting from an extremely thin layer of glass. To draw the front reflection arrow, we imagine a photon leaving the light source (the stopwatch hand starts turning), bouncing off the front surface, and arriving at A (the stopwatch hand stops). We draw a little arrow of length 0.2 in the direction opposite that of the stopwatch hand (see Fig. 10).

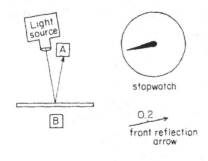

FIGURE 10. *In an experiment measuring reflection by two surfaces, we can say that a single photon can arrive at A in two ways—via the front or back surface. An arrow of length 0.2 is drawn for each way, with its direction determined by the hand of a "stopwatch" that times the photon as it moves. The "front reflection" arrow is drawn in the direction opposite to that of the stopwatch hand when it stops turning.*

To draw the back reflection arrow, we imagine a photon leaving the light source (the stopwatch hand starts turning), going through the front surface and bouncing off the back surface, and arriving at A (the stopwatch hand stops). This time, the stopwatch hand is pointing in almost the same direction, because a photon bouncing off the back surface of the glass takes only slightly longer to get to A—it goes through the extremely thin layer of glass twice. We now draw a little arrow of length 0.2 in the same direction that the stopwatch hand is pointing (see Fig. 11).

Now let's combine the two arrows. Since they are both the same length but pointing in nearly opposite directions, the final arrow has a length of nearly zero, and its square is even closer to zero. Thus, the probability of light reflecting from an infinitesimally thin layer of glass is essentially zero (see Fig. 12).

FIGURE 11. *A photon bouncing off the back surface of a thin layer of glass takes slightly longer to get to A. Thus, the stopwatch hand ends up in a slightly different direction than it did when it timed the front reflection photon. The "back reflection" arrow is drawn in the* same *direction as the stopwatch hand.*

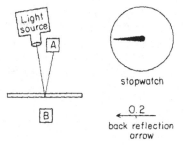

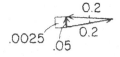

FIGURE 12. *The final arrow, whose square represents the probability of reflection by an extremely thin layer of glass, is drawn by adding the front reflection arrow and the back reflection arrow. The result is nearly zero.*

When we replace the thinnest layer of glass with a slightly thicker one, the photon bouncing off the back surface takes a little bit longer to get to A than in the first example; the stopwatch hand therefore turns a little bit more before it stops, and the back reflection arrow ends up in a slightly greater angle relative to the front reflection arrow. The final arrow is a little bit longer, and its square is correspondingly larger (see Fig. 13).

As another example, let's look at the case where the glass is just thick enough that the stopwatch hand makes an extra half turn as it times a photon bouncing off the back surface. This time, the back reflection arrow ends up pointing in exactly the same direction as the front reflection arrow. When we combine the two arrows, we get a final arrow whose length is 0.4, and whose square is 0.16, representing a probability of 16% (see Fig. 14).

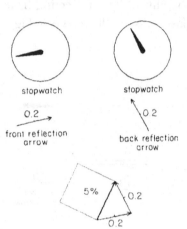

FIGURE 13. *The final arrow for a slightly thicker sheet of glass is a little longer, due to the greater relative angle between the front and back reflection arrows. This is because a photon bouncing off the back surface takes a little longer to reach A, compared to the previous example.*

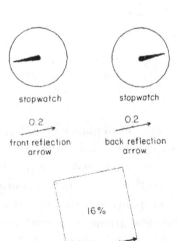

FIGURE 14. *When the layer of glass is just thick enough to allow the stopwatch hand timing the back reflecting photon to make an extra half turn, the front and back reflection arrows end up pointing in the same direction, resulting in a final arrow of length 0.4, which represents a probability of 16%.*

If we increase the thickness of the glass just enough sothat the stopwatch hand timing the back surface path makes an extra *full* turn, our two arrows end up pointing in opposite directions again, and the final arrow will be zero (see Fig. 15). This situation occurs over and over, whenever the thickness of the glass is just enough to let the stopwatch hand timing the back surface reflection make another full turn.

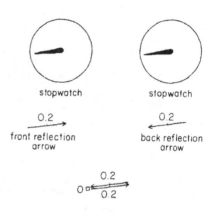

FIGURE 15. *When the sheet of glass is just the right thickness to allow the stopwatch hand timing the back reflecting photon to make one or more extra full turns, the final arrow is again zero, and there is no reflection at all.*

If the thickness of the glass is just enough to let the stopwatch hand timing the back surface reflection make an extra ¼ or ¾ of a turn, the two arrows will end up at right angles. The final arrow in this case is the hypoteneuse of a right triangle, and according to Pythagoras, the square on the hypoteneuse is equal to the sum of the squares on the other two sides. Here is the value that's right "twice a day" - 4% + 4% makes 8% (see Fig. 16).

Notice that as we gradually increase the thickness of the glass, the front reflection arrow always points in the same direction, whereas the back reflection arrow gradually changes its direction. The change in the relative direction of the two arrows makes the final arrow go through a repeating cycle of length zero to 0.4; thus the *square* on the final arrow goes through the repeating cycle of zero to 16% that we observed in our experiments (see Fig. 17).

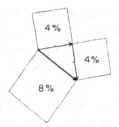

FIGURE 16. *When the front and back reflection arrows are at right angles to each other, the final arrow is the hypoteneuse of a right triangle. Thus its square is the sum of the other two squares—8%.*

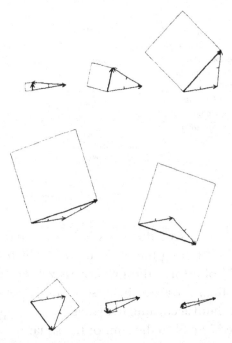

FIGURE 17. *As thin sheets of glass are replaced by slightly thicker ones, the stopwatch hand timing a photon reflecting off the back surface turns slightly more, and the relative angle between the front and back reflection arrows changes. This causes the final arrow to change in length, and its square to change in size from 0 to 16% back to 0, over and over.*

I have just shown you how this strange feature of partial reflection can be accurately calculated by drawing some damned little arrows on a piece of paper. The technical word for these arrows is "probability amplitudes," and I feel more dignified when I say we are "computing the probability amplitude for an event." I prefer, though, to be more honest, and say that we are trying to find the arrow whose square represents the probability of something happening.

Before I finish this first lecture, I would like to tell you about the colors you see on soap bubbles. Or better, if your car leaks oil into a mud puddle, when you look at the brownish oil in that dirty mud puddle, you see beautiful colors on the surface. The thin film of oil floating on the mud puddle is something like a very thin sheet of glass—it reflects light of one color from zero to a maximum, depending on its thickness. If we shine pure red light on the film of oil, we see splotches of red light separated by narrow bands of black (where there's no reflection) because the oil film's thickness is not exactly uniform. If we shine pure blue light on the oil film, we see splotches of blue light separated by narrow bands of black. If we shine both red *and* blue light onto the oil, we see areas that have just the right thickness to strongly reflect only red light, other areas of the right thickness to reflect only blue light; still other areas have a thickness that strongly reflects both red and blue light (which our eyes see as violet), while other areas have the exact thickness to cancel out all reflection, and appear black.

To understand this better, we need to know that the cycle of zero to 16% partial reflection by two surfaces repeats more quickly for blue light than for red light. Thus at certain thicknesses, one or the other or both colors are strongly reflected, while at other thicknesses, reflection of both colors is cancelled out (see Fig. 18). The cycles of reflection repeat at different rates because the stopwatch hand turns around faster when it times a blue photon thanit does when timing a red photon. In fact, that's the only difference between a red photon and a blue photon (or a photon of any other color, including radio waves, X-rays, and so on)—the speed of the stopwatch hand.

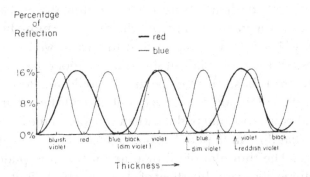

FIGURE 18. *As the thickness of a layer increases, the two surfaces produce a partial reflection of monochromatic light whose probability fluctuates in a cycle from 0% to 16%. Since the speed of the imaginary stopwatch hand is different for different colors of light, the cycle repeats itself at different rates. Thus when two colors such as pure red and pure blue are aimed at the layer, a given thickness will reflect only red, only blue, both red and blue in different proportions (which produce various hues of violet), or neither color (black). If the layer is of varying thicknesses, such as a drop of oil spreading out on a mud puddle, all of the combinations will occur. In sunlight, which consists of all colors, all sorts of combinations occur, which produce lots of colors.*

When we shine red and blue light on a film of oil, patterns of red, blue, and violet appear, separated by borders of black. When sunlight, which contains red, yellow, green, and blue light, shines on a mud puddle with oil on it, the areas that strongly reflect each of those colors overlap and produce all kinds of combinations which our eyes see as different colors. As the oil film spreads out and moves over the surface of the water, changing its thickness in various locations, the patterns of color constantly change. (If, on the other hand, you were to look at the same mud puddle at night with one of those sodium streetlights shining on it, you would see only yellowish bands separated by black—because those particular streetlights emit light of only one color.)

This phenomenon of colors produced by the partial reflection of white light by two surfaces is called iridescence, and can be found in many places. Perhaps you have wondered how the brilliant colors of hummingbirds and peacocks are produced. Now you know. How those brilliant colors evolved is also an interesting question. When we admire a peacock, we should give credit to the generations of lackluster females for being selective about their mates. (Man got into the act later and streamlined the selection process in peacocks.)

In the next lecture I will show you how this absurd process of combining little arrows computes the right answer for those other phenomena you are familiar with: light travels in straight lines; it reflects off a mirror at the same angle that it came in ("the angle of incidence is equal to the angle of reflection"); a lens focuses light, and so on. This new framework will describe everything you know about light.

INDEX